AF347521

Sustainable Built Environments in 21st Century

Sustainable Built Environments in 21st Century

Editors

Elmira Jamei
Zora Vrcelj

MDPI • Basel • Beijing • Wuhan • Barcelona • Belgrade • Manchester • Tokyo • Cluj • Tianjin

Editors
Elmira Jamei
Victoria University
Australia

Zora Vrcelj
Victoria University
Australia

Editorial Office
MDPI
St. Alban-Anlage 66
4052 Basel, Switzerland

This is a reprint of articles from the Special Issue published online in the open access journal *Applied Sciences* (ISSN 2076-3417) (available at: http://www.mdpi.com).

For citation purposes, cite each article independently as indicated on the article page online and as indicated below:

LastName, A.A.; LastName, B.B.; LastName, C.C. Article Title. *Journal Name* **Year**, *Volume Number*, Page Range.

ISBN 978-3-0365-2848-9 (Hbk)
ISBN 978-3-0365-2849-6 (PDF)

Cover image courtesy of New York Times

Contents

About the Editors

Elmira Jamei is a Senior Lecturer and course chair of the Building Design program. She is also the deputy program leader for the Infrastructure and the Built Environment program at the Institute for Sustainable Industries and Liveable Cities at VU.

Before being appointed to Victoria University, she contributed to teaching and research in Architecture and Urbanism across Australia and Asia, where she also practised as an architect and urban designer in diverse design companies. Elmira has worked for 10 years as the project leader and key researcher in several national and international collaborative projects within Asia, Australia, and the Middle East.

She led several successful grants, collaborative research projects and industry-engaged works, which resulted in publications in high-impact research journals and awards, including 'Vice Chancellor's Citation Award for Excellence in Research and Research Training', 'Graham Treloar Research Prize', and 'Collaborative research prize' from Chinese Academy of Science.

Elmira brings a substantial knowledge base and practical skills in environmentally sustainable design, smart cities, climate-sensitive design, renewable energies in architecture and planning, and the use of novel visualization techniques in built environment. She has led and coordinated a number of projects in the area of energy, urban thermal balance and performance of buildings in diverse climates.

Zora Vrcelj As Head of Built Environment at Victoria University, Associate Professor Zora Vrcelj enjoys working across the built environment, engineering and construction sector to achieve impactful outcomes.

Prior to joining Victoria University in 2012, Associate Professor Vrcelj was an academic for 12 years within the School of Civil and Environmental Engineering at the University of New South Wales (UNSW).

Associate Professor Vrcelj has been working to realise technological developments aimed at enhancing the functionalities of high-performance structures (smart buildings and civil infrastructure). She has worked on numerous research projects, with particular focus on structural engineering design, advanced composite structures embedded with piezoelectric sensors and actuators, and testing and evaluation of materials used in buildings and civil infrastructure. This experience, combined with her work in the fields of structural engineering, structural mechanics, materials science and smart buildings, enables her to extrapolate this expertise to built environment practice.

Associate Professor Vrcelj has published over 80 refereed technical publications and has secured numerous competitive grants (including ARC DP and CRC grants). Since 2010, she has held a Visiting Professor position in Civil engineering at the University of Novi Sad, Serbia

Preface to "Sustainable Built Environments in 21st Century"

The built environment is where we live and spend most of our time. It is a crucial sector for the economy, accounting for roughly 10% of global GDP, but it is also a key driver of emissions, resource use and land use change across multiple sectors. In fact, it represents over one-third of global final energy use, generates nearly 40% of global energy-related greenhouse gas (GHG) emissions, and consumes almost half of the total global raw material. The world will need an extensive expansion of urban infrastructure in the coming decades due to rapidly growing cities in emerging economies and new family structures and aging populations in developed economies. To welcome future space demands and limit global warming, we need to accelerate the shift toward net-zero emissions and a circular, healthy, inclusive, and sustainable built environment.

Sustainable built environments aim to achieve unprecedented levels of ecological balance through new and retrofit construction and built environment, especially in the 21st century when rapid level of urbanization and population growth are inevitable. To create spaces with both long-term viability and humanization, we merge the natural, minimum resource conditioning solutions of the past with the innovative technologies of the present and consider both environmental and social impacts. This may include land use, biodiversity damages, urban climate, energy consumption, carbon emission, material, water consumption, social inclusion, public health, and other impacts.

The focus of this Special Issue will be studies related to the innovative approaches and solutions of sustainable built environments, including design, construction, urban design, assessment models, case studies, and best practice analyses or their limitations.

Elmira Jamei, Zora Vrcelj
Editors

Review

Biomimicry and the Built Environment, Learning from Nature's Solutions

Elmira Jamei * and Zora Vrcelj

College of Engineering and Science, Victoria University, Melbourne, VIC 3011, Australia; zora.vrcelj@vu.edu.au
* Correspondence: elmira.jamei@vu.edu

Abstract: The growing interest in biomimicry in built environments highlights the awareness raised among designers on the potentials nature offers to human and system function improvements. Biomimicry has been widely utilized in advanced material technology. However, its potential in sustainable architecture and construction has yet to be discussed in depth. Thus, this study offers a comprehensive review of the use of biomimicry in architecture and structural engineering. It also reviews the methods in which biomimicry assists in achieving efficient, sustainable built environments. The first part of this review paper introduces the concept of biomimicry historically and practically, discusses the use of biomimicry in design and architecture, provides a comprehensive overview of the potential and benefits of biomimicry in architecture, and explores how biomimicry can be utilized in building envelops. Then, in the second part, the integration of biomimicry in structural engineering and construction is thoroughly explained through several case studies. Finally, biomimicry in architectural and structural design of built environments in creating climate-sensitive and energy-efficient design is explained.

Keywords: biomimicry; architecture; structural engineering; sustainable design

Citation: Jamei, E.; Vrcelj, Z. Biomimicry and the Built Environment, Learning from Nature's Solutions. *Appl. Sci.* **2021**, *11*, 7514. https://doi.org/10.3390/app11167514

Academic Editor: Pietro Picuno

Received: 10 May 2021
Accepted: 10 August 2021
Published: 16 August 2021

Publisher's Note: MDPI stays neutral with regard to jurisdictional claims in published maps and institutional affiliations.

1. Introduction

Biomimicry is the design that is inspired by nature in terms of functional concepts of an organism or an ecosystem [1]. According to Janine Benyus, bio-mimicry mimics processes in nature to create innovative and sustainable design solutions [2]. She also describes biomimicry as a science in which nature is considered the mentor and model for design [2,3]. In general, biomimicry uses ecological benchmarks to assess sustainability and create vernacular designs inspired by nature in terms of form, process, and ecosystems [2]. Other scholars have perceived biomimicry as a field of science that aims to address human needs through mimicking natural designs, processes, and systems [4,5]. Biomimicry is a multidisciplinary field of research where experts with diverse backgrounds (e.g., philosophy, computer science, physics, and chemistry) work together with biologists and engineers to create highly resilient products. Biomimicry is quite critical for today's world, where rapid climate change and environmental degradations occur.

Historically, the art of biomimicry goes back to 500 B.C., when Greek philosophers learned from the natural organisms and applied their mechanisms, shapes, and functions as the model to make the balance between different parts of design and create the classical idea of beauty [6]. Later, in 1482, Leonardo Da Vinci invented the flying machine by studying the mechanism of birds flying and labeled his work as the early example of biomimicry [6]. Although he was unsuccessful with the flying machine, his invention later led to the development of Wright's brother's prototype to an airplane in 1948 [7]. In 1958, the term bionics was first introduced and defined as 'the science of natural systems or their analogs'. However, the term biomimicry did not appear before 1982. Later, in 1997, Janine Benyus expanded the concept of biomimicry in her book 'Biomimicry: Innovation inspired by Nature'. Then, she established the Biomimicry Institute with Schwan. In 2007, Chris

Allen joined the company to launch 'Ask Nature', known as the world's first digital library, which gives natural solutions and inspirations to design practice and research.

Biomimicry is different from bionic. Bionics is the design of engineering systems, especially electronic ones, based on biological systems, whereas biomimetic is the study of the structure and function of living things as models for creating materials or products by reverse engineering [8]. The act of studying and mimicking nature to come out with practical solutions that address human needs is not a novel practice. In the past, people were often inspired by nature to provide their food, shelter, and innovative methods to survive in harsh environments. These innovative methods have been re-used in the contemporary era in the fields of built environment, medical science, defense, agriculture, and even manufacturing processes [8–10].

The ecosystem and nature can be mimicked and contribute to the resilient, sustainable, and adaptable built environment, which improves the capacity of regeneration in the natural environment and adaptation against climate change [11]. Biomimicry also offers thoughtful solutions for human needs through a translational process into a human context where the design may not be similar to the source organism/ecosystem but poses the same functional concepts.

Early scientists have conducted in-depth studies on the functions and processes in nature. They have collected valuable information used in different areas of study, particularly design, architecture, and structural engineering. Thus, this study aims to review the use of biomimicry in architecture and structural engineering and investigates how biomimicry contributes to a sustainable and resilient built environment.

2. Biomimicry in Architecture

2.1. Concept of Biomimicry in Design and Architecture

According to Feuerstein and Fred Otto [12,13] biology and architecture are prerequisites of each other. Bioinspiration in architecture is understood as a practical methodology for answering the stakes of designs of forms and energy-efficient structures at the urban scale using natural materials. Biomimetic architecture aims to measure and shape space and to create synergistic relations between the environment and the structure.

The adaptability of nature toward different environmental changes has been well reported in the literature. This adaptability of nature has inspired several designers to create highly resilient and environmentally sustainable built environments [14]. This inspiration from nature has evolved in two ways in the context of design and architecture: direct and indirect approaches. Scholars [15–18] have comprehensively studied the features and characteristics of each approach in the work of well-known architects and designers.

The direct design approach occurs when a design directly copies an organism in the ecosystem and mimics its behavioral pattern or natural system. Whereas, the indirect approach solely uses abstract concepts in nature and employs them in design [19,20]. The direct design approach has two derivations with two diverse schools of thought and methods. The first approach understands the design problems based on a 'design exploring biology' concept, and the second approach explores the design issues from a 'biology investigating design' perspective [15]. The latter consists of identifying the human needs or design issues through understanding the processes that the ecosystems utilize to overcome such challenges.

In architecture, Biomimicry is also known for its problem-driven or solution-driven approach to architectural design issues [17]. In this approach, the designer explores solutions to address the problems through biology, whereas in the solution-driven approach biology is used as a solution to copy and then transfer to design systems.

Biomimicry inspires architecture in three ways; organism (imitation of nature), behavior (imitation of natural processes), and ecosystem levels (imitation of the working principles of ecosystems) [15]. At the organism level, design and architecture are mainly inspired by the form, shape, or structure of a building. At the behavioral level, the interaction between the ecosystem and its surroundings inspires the design. At the ecosystem level, the main focus is on how different parts of an organism interact on a large (urban) scale. Table 1 summarises the characteristics of each level. This approach has been methodized to apply to a design or an architectural problem [21].

Table 1. Framework for different levels of biomimicry.

Organism Level	Behaviour Level	Ecosystem Level
Mimicry of a specific organism	Mimicry of the way that organs behave of a larger context	Mimicry of an ecosystem

These levels have been thoroughly explained by Benyus through an example of an owl's feather. A feather can be renewed by its formal attributes. However, this replication cannot be considered a resilient and sustainable solution [2]. When the process is mimicked, identifying how the feather is produced without using toxic waste or a high level of energy consumption is feasible—realizing that how it impacts body heat and energy conservation and thereby achieves the properties of the feather is possible. At the ecosystem level, the existence of the bird and its feather with a larger biosphere and the entire organism is studied.

Each of these levels offers five potential dimensions to biomimicry: (1) how the design mimics the look and form of an ecosystem, (2) how it mimics the material of an ecosystem, (3) how it mimics the way that the ecosystem is being constructed, (4) how the ecosystem works (process) and (5) what the ecosystem is capable of doing (function). These levels are often used as benchmarks for architects to employ bio-mimicry principles in design and architecture approaches and create sustainable, efficient, and environmentally sound buildings.

In some architecture and design concepts, most projects are inspired by the form and behavior of certain animals (animals in the ocean or on the earth) that have adaptive approaches towards the outside world (e.g., sun and wind). In other architectural projects, the source of inspiration is plants that react differently towards extreme climatic conditions (drought, heat, and light).

2.2. Potentials and Benefits of Biomimicry in Architecture and Design

As discussed, biomimicry brings several inspirations from nature and introduces great potentials to create a sustainable, energy-efficient built environment. This great opportunity is more tangible, particularly today, because new building materials and new construction techniques can be seen more than ever in the past. However, the method in which the built environment reaches its final form is crucial. Therefore, the significance of a well-designed built environment lies in integrating creative processes learned from nature (biomimicry) and the wealth of knowledge in technology and tools. Therefore, the next sections of this paper provide an overview of the potential use of biomimicry in architecture.

2.2.1. Wise Selection of Construction Materials
Function

Wise material selection with a high level of functionality is one of the major benefits of applying biomimicry in architecture. The importance of understanding complex systems results in consideration of the individual aspects, which leads to an improved understanding of the overall function.

In nature, efficient materials are defined as those that have effective exchange with expensive materials (which are generated from metabolic processes). Nature has created sustainable light shell and fold structures and systems that can grow and be stable. Natural systems established the building processes in both animals and plants. This building process considers the availability of local materials and aims to create an optimized and multi-functional structure. Examples of such building processes can be seen in shell structures of mussels and sea and folded structures of leaves, hornbeam, and palm varieties.

Lifecycle

The life cycle is another great lesson learned from nature and be is implemented in architecture, whether the matured structures are occupied by new life forms or decomposed into basic elements, from which new life forms can emerge. Biomimicry in architecture has resulted in building materials and elements that can integrate themselves with a life cycle in nature. However, a tangible gap has been observed in the literature on how the life cycle of built environments can learn lessons from the natural processes and ecosystems in nature [22,23].

Weight

The concept of lightweight structures is another potential brought by biomimicry in architecture and building methods. Natural structures react to internal and external loads differently. Thus, their forms are affected by such factors, which is also the case for human-made technology-driven built environments. One of the benefits of using lightweight materials for building envelopes is the high level of insulation and light penetration, diffusion. An example of these features in nature can be seen in polar bear fur. It provides good insulation for the cold weather of Antarctica and allows the penetration of light into the darkly pigmented skin of the bear [24]. Another similarity can be seen between the hairs and parallel glass fibers, acting as the insulator and light distributor [25].

2.2.2. Structure Behaviour

The possibility of creating an evolutionary and evolving urban planning and design is another inspiration brought by biomimicry in architecture and design. The opportunity of using advanced and technology-driven tools enables the designers to choose the processes that are similar to those in nature and ecosystems. One of the pioneering institutions in using computer-/technology-supported algorithms for evolutionary urban design and urban planning is the Institute for Computer-Based Design at the University of Stuttgart in Germany. In one of the projects conducted by this institution, the structure of a building was thoroughly analyzed and optimized to improve the Structure's behavior, function, and, in certain circumstances, its mobility [26].

2.2.3. Building Envelope (Heating, Cooling, and Lighting)

There are infinite sources of inspiration from nature that can be utilized in different design and construction technologies and contribute to the effective algorithm, method, material, processes, structure, tool, mechanism, and systems. Living organisms have unique integration geometries and techniques that enable them to adapt themselves to harsh-diverse environments easily. Similarly, buildings nowadays use specific methods to adapt well to their surrounding environments and minimize the adverse impact on the environment.

Designing the building envelope is among the important methods. The building envelope, also known as the third skin, is 'an extended buffer between the building and the exterior environment'. The first human shelters and settlements consisted of cloths or natural caves. Later, these shelters were built with raw materials, and nowadays, we see communities where houses are built to protect one another and thus create a single unit with an external wall. However, with the development of individual buildings, the optimized envelop design and multi-layered construction have increased. Past services were mainly attached to the envelope and provided isolated solutions while neglecting the building features and improving the constant need for maintenance.

2.2.4. Building Facades

The biomimicry approach is not about copying nature in form but also learning from its principles and methods and coming up with sophisticated technological solutions for efficient building envelopes. One of these technologies is the techniques applied in building skin.

Building skin is a thin membrane that comes on top of the structure and regulates the mechanical and electrical function of the structure that also forms the buildings' interior spaces. There is a similarity between building skin as what we know as façade and natural skin in nature. Both consist of diverse layers that filter external newcomers and react differently to heat, pollution, water, and noise pollution. One of the main overlaps between these two types of skins is that both keep the condition of internal spaces constant while meeting the functional need of the space. They both act as a filter in the process of determining what is allowed to enter and exit [27].

The main benefit of utilizing biomimicry is that designing building skins creates an efficient thermoregulatory mechanism (such as heating, cooling, and lighting). To create a nexus between building skin and biomimicry, we need first to analyze the commonalities between the living ecosystems and building facades and the driving forces that influence the nature and design process. One of these similarities is the tendency of living organisms to adapt their temperature to their surrounding environments and maintain a steady condition. Similarly, animals constantly modify their structures and behaviors to maximize the use of available accessible sources of energy (e.g., wind, sun, and water).

2.2.5. Heating and Insulation

Similar to built environments, in ecosystems, heat is transferred through radiation, evaporation, conduction, and convection. In some animals, heat is generated inside the body, and the body then tends to keep the temperature steady. Whereas, in other types, heat is mainly absorbed from the environment, and the body temperature ranges quite significantly. The first type of animal and the concept of generating the heat from metabolism has been the idea behind heating techniques in many buildings. In this type of buildings, the spaces are kept warm by preventing heat loss. Therefore, insulation plays a critical role in addressing this objective. Polar bears in the Antarctic, and their bodies are the best examples of such adaptation capabilities. Layers of fat and a denser layer of fur act as insulation. Their hollow hair fiber adds to insulation strength. Similarly, in other animals, hair filaments conduct sunlight down to their dark skin to create a curtain wall system that automatically modifies their insulations.

2.2.6. Direct Heat Gain

Another method in keeping the space warm is through direct heat gain from the sun. Communal nests built from the silk layers that are oriented towards the southeast to capture the heat from the sun are a good example of this method. In this method combined effect of insulation and solar orientation can lead to a 4 °C higher temperature [28].

An example is how penguins create heat. Penguins live in groups, and their skins have a constant temperature, regardless of the ambient temperature around them. Penguins huddle in groups to reduce the exposure of their surfaces towards outside areas. A similar design principle can be seen in vernacular architecture where buildings link to each other, and the only open space is the atrium-shaped opening, which is mainly used for ventilation purposes (and often closed in winter to decrease heat loss).

Reducing heat loss from buildings would result in warmer indoor temperatures. Several passive Haus projects have adopted this concept where the heating system is mainly relying on the internal heat gains obtained from the metabolisms of the occupants and equipment in the building. One of the built examples of this biomimetic principle can be seen in the Himalayan rhubarb towers, where a vertical greenhouse of translucent leaves contributed to a 10 °C higher temperature in indoor spaces compared with outdoor ambient air temperature [29] to balance heat loss through the skin.

2.2.7. Cooling

Some living organisms that live in extremely hot regions avoid radiative heat gain by staying out from the sun or relieving from conductive heat gain by minimizing their skin exposure to the sun (skipping across the sand). This principle (avoiding direct heat gain) has become the main action plan in architecture for cooling buildings. This principle seems straightforward. However, its importance has not been highlighted till the late 20th century. A similar approach in architecture and design can be found in Cabo Llanos Tower in Santa Cruz de Tenerife, Spain by Foreign Office Architects and the Singapore Arts Centre by Michael Wilford and Partners with Atelier One and Atelier Ten

Another example is the work of Chuck Hoberman [30], who is also one of the pioneers in adaptive approaches towards solar shading. One example where a shading device is integrated into the building body is Hoberman's dynamic windows for The State University of New York's Simon Centre for Geometry and Physics (Figure 1). The windows function as the artistic centerpiece of the building and the functional shading piece. Every project panel is created for a distinctive geometric perforation pattern mirroring building resident mathematicians and scientists' research focus. The patterns range in line and diverge. Some geometric patterns with circles, hexagons, triangles, and squares are seen flourishing into an opaque mesh, and thereby lead to a higher level of control over the received sunlight.

Figure 1. An example of shading devices used as an essential part of a structure for controlling solar radiation is the dynamic windows at the Simon Centre for Geometry and Physics at the State University of New York, USA (2010) (Source: https://www.hoberman.com/portfolio/dynamic-windows/).

South Korea's Thematic Pavilion is also an example of an integrated self-shaded device. It is inspired by a South African flower and has a movable 90-degree flap. The main use of the principle is solar shading with minimized view obstruction when cloudy weather and full protection from the sun when sunny [31]. The Pavilion is based on Strelitzia Reginae's movements. The perch curves and the petals open whenever a bird lands on the flower, revealing the anther to the bird and making pollination possible. Researchers at the University of Freiburg's Plant Biomechanics Group used this concept for shading. They later designed a shading principle wherein shading is available when needed and can be moved away when not, preventing view obstruction. The Pavilion's shading method minimizes the sun's radiative heat using 108 kinetic lamellas. The glass-fiber-reinforced polymers are used to make the lamellas for low bending stiffness and high tensile strength, allowing for reversible deformations. This principle was needed in adjusting the lamellas' bending to control solar input. The solar panels on the rooftop charge the actuators. Similar to an anther moving in and out during pollination by the bird, the lamellas twist to control the solar gain.

2.2.8. Thermoregulation

A critical mechanism in cooling the building is efficient thermoregulation. One of the manifestations of inspiration from termite mounds in thermoregulation (for cooling purposes) is Western Australia's mounds caused by compass termites [32]. The compass termites form an almond-shaped plan with a long axis oriented towards the north and south. The heat from the morning sun is absorbed through flat sides, and the mid-day heat is least absorbed by minimizing the exposure area (Figure 2). Termite also controls ventilation tubes. The rising inside temperature increases, opens the ventilation tubes, and lets the heat rise through a stack effect.

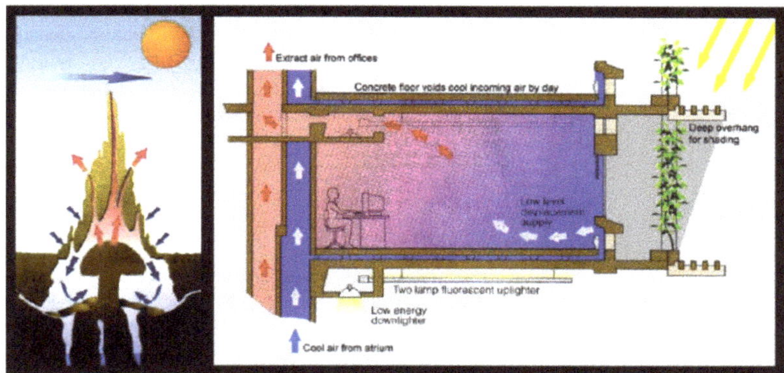

Figure 2. Section of a termite-inspired building that can cool itself (**Left**), Heat circulation in a room and a termite hill (**Right**) (Source: https://parametrichouse.com/biomimicry-architecture-2/).

The office buildings and shopping complexes in the Eastgate Centre have stable air temperature indoors all year. The center does not use mechanical cooling or heating system and consumes only 10% of the energy used in a conventional structure. Its porosity (Figure 3) causes the vents to pull in air, which cools as it enters the building because of heat-absorbing concrete slabs. The center's system is highly effective because the accumulated heat is sent to the slabs. Losing or gaining air depends on whether the concrete or air is cool. The air moves into the occupied spaces, then rises and flows up through exhaust. The released cycle draws through the Structure, consistently circulating fresh air.

Figure 3. The Eastgate Centre has mimicked termite performance in passive cooling through the use of local material and porous building.

The building's self-contained system is used for night ventilation. The high-volume fans move at a rate of 10 air changes/hour. The air goes into occupied spaces through centralized ducts. The air travels via hollow floors and is released from the low-level window grills [33].

The design concept integrates the regionalized stone style and the international glass and steel style. The building's cooling system is inspired by the local termite hills' passive cooling. The local biological system's design made the mimicry environmentally conducive and provides a network that sustains comfortable temperature even without a heating, ventilation, and air conditioning (HVAC) system. To ensure termite survival, the hill's internal temperature must be sustained at a constant temperature of 30.6 °C, but its external temperature may vary between 1 and 40 °C. Termites constantly adjust the air sucked through the mounts to ensure the survival of the fungus they consume. The vents are adjusted to open or close, depending on the required changes. The surrounding clay absorbs the heat and cools the air. The warm air in the mound rises through the central ventilator, releasing hot air to outdoor spaces and absorbing the cool air.

2.2.9. Lighting

Lighting has a well-established impact on human wellbeing and lifestyle. Tado Ando and Le Corbusier highlighted the fundamental role lighting plays in buildings and how it impacts us in three ways: radiation, our visual systems, and our circadian system [34].

Biomimicry offers diverse potential solutions and inspirations for designing lighting in architectural projects. Nature takes two aspects of light and color. Therefore, lighting must be considered when designing a biomimicry-inspired project.

One of the biomimetic design concepts which can be used in lighting design is gathering and focusing the light. For example, an anthurium offers some interesting aspects for collecting light in diffused conditions. Similarly, a spookfish inspires the idea of integrating a symmetrical pair of mirrors in the atrium spaces to reflect the light into building interior spaces [35].

For example, Pawlyn's recent project, The Biomimetic Office, is inspired by the spookfish's (Figures 4 and 5) [1] way of focusing low light levels. Architects can emulate the spookfish's ability when designing buildings. At first, this vertebrae spookfish was believed to have four eyes but was later found to utilize mirrors instead of lenses to focus light with eyes. Each eye has two connected parts. One points upwards and towards daylight, whereas the other points downwards. A mirror is used for focusing low-intensity light from bioluminescence. Pawlyn uses the spookfish's mirroring method to disperse natural light in his building, reduce energy use and raise occupants' wellbeing.

The angled plates of the mirror in the spookfish eye create a curved shape that allows the maximum amount of reflected light and the sharpest possible image (Figure 5a,b). The fish is predicted to change the mirror's position to center on objects from varied distances.

Minimizing the self-shading through the building is another concept widely used in lighting design. Another biomimetic design concept in lighting is minimizing the self-shading through the building itself. This principle is mainly seen among plants with phyllo-tactic geometry is often employed in the lighting design of buildings. Their form deeply harnesses the light. These projects used the Fibonacci rule on the ratio of series of repeating spirals.

In [1], the architect also proposed phyllotactic towers that act as a private garden for each housing and maximize solar heat gain and energy harvesting opportunities. Saleh Masoumi, the architect of Verk Studio in Iran, offered a novel solution to residential towers. His designs are inspired by the structure of living plants, providing each residential/work unit with 'yards'. In botany, phyllotaxis or basic leaf patterns could be alternating or opposite around the plant's stem.

Figure 4. Exploration Architecture's Biomimetic Office mimics the spookfish's eye structure (https://www.dezeen.com/2020/10/22/michael-pawlyn-exploration-architecture-dassault-systemes-video/, accessed on 22 October 2010).

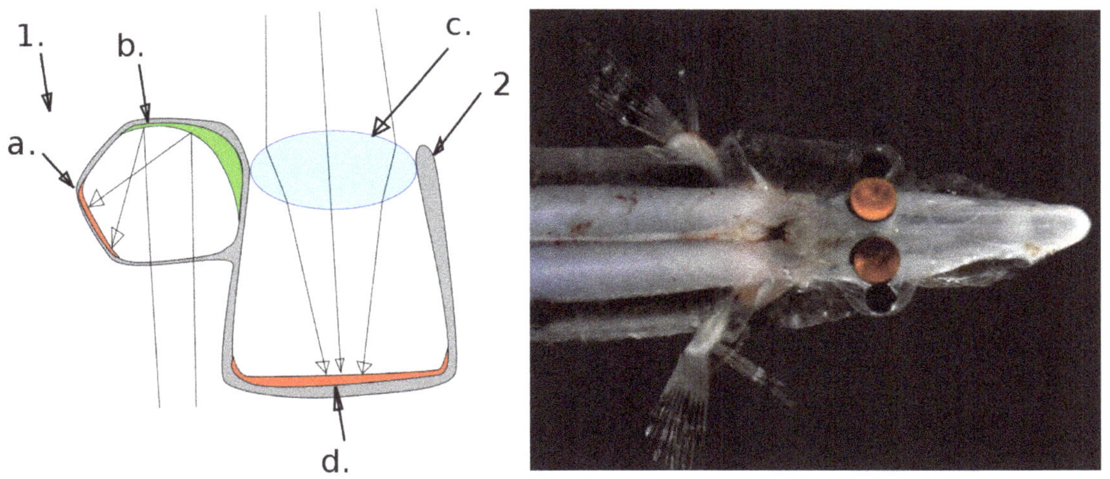

Figure 5. The downward-facing eye of the brownsnout spookfish (1) centers light on the retina (**a**) using reflective crystals (**b**). The main eye (2) uses a lens (**c**) and center light on the retina (**d**) (Left). The brownsnout spookfish's additional eye structures let the fish see below. (Credit: Florida Atlantic University (Right, (Source: https://asknature.org/strategy/extra-eyes-direct-light/).

3. Biomimicry in Structural Engineering

The age of industrial evolution devised the divergence of humanity from nature [36]. However, engineers almost always constructed structures and machinery using the 'heat, beat and treat' principle [2] by applying large amounts of heat, large pressure, and various toxic chemical treatments. A rapid increase in greenhouse emissions and carbon dioxide in urban areas has led to serious environmental degradation and posed great risks for public health. The construction boom and built environment are known as major contributors in accelerating these degradations and generating a high level of pollution and energy demand [37]. Furthermore, products developed by humans often cannot be recycled, thus polluting the planet utilizing land waste.

Although biomimicry has attracted reasonable attention in the fields of mechanical engineering (robotics), materials science (intelligent materials), and biomedical engineering (prosthetics), it remains a grey area in structural engineering. Engineers and environmental scientists have attempted to mimic forms and designs of nature to apply findings to practical structural engineering problems and achieve reasonable solutions (higher strength or fewer resources required) that address environmental and sustainability issues. Imitating shapes and geometry of structures from nature is the best-known biomimicry in structural engineering. For example, the roof of Pantheon in Rome gains its strength from its multi-dimensional curvature by mimicking the shape of a seashell, resulting in lightweight and reduced reinforcement [38].

By studying how natural structures/systems sustain loads and optimize resources existing structural design strategies can be improved or reinvented to achieve efficient and sustainable built environments. Sustainable interferences are needed while creating these built environments and not after building them [37].

In addition to studying the forms and designs of nature, imitating the natural processes is another promising avenue for adopting biomimicry to construct contemporary built environments. Superstructures, such as dams, have been built to generate power for human activities, divert and supply water for agriculture, prevent flooding and stabilize the water. Although hydropower is considered green energy, greenhouse gases have been generated by constructing dams. Beavers create dams by piling up twigs, branches, and trunks of trees (Figure 6). The construction process of beavers' dams reveals the acquisition and utilization of local materials, the choice of reusable and recycled materials, and the increased efficiency of the system.

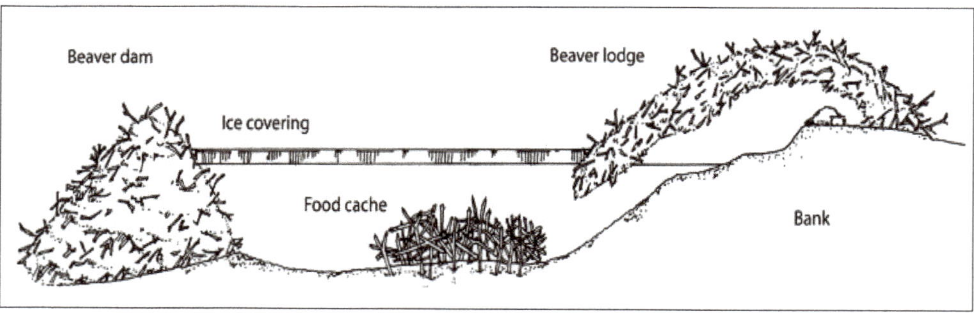

Figure 6. Diagram of beaver's dam and lodge (Source: https://www.hww.ca/en/wildlife/mammals/beaver.html).

The current construction process is typically powered by renewable energy, such as chemical energy or sunlight. As a result, if scientists and engineers can crack the secret of 'natural construction', our manufacturing and construction processes will face a breakthrough. The manufacturing process no longer requires enormous energy input, which can significantly reduce cost and pollution. Engineers and scientists have many future possibilities in mimicking nature and developing engineering designs and construction.

However, we must first unravel and understand the basic principles. A fundamental point of biomimetic is to understand the principles and the reasons why things work in nature. According to Biomimicry 3.8 (2021), the six major biomimicry principles are as follows:

(i)   Resource (material and energy) efficiency
(ii)   Evolution for survival
(iii)   Adapting to changing conditions
(iv)   Integrating development with growth
(v)   Being locally attuned and responsive
(vi)   Using life-friendly chemistry

Given that structural engineers' knowledge of biology is limited, there is a need to raise awareness of these principles and their use in structural engineering.

3.1. Concept of Biomimetics and Structural Engineering Design Process

Biomimetics benefits the structural engineering design process. In structural engineering, the main objective is to design structures to achieve functionality and maintain their structural integrity during the design life. For example, structural engineers design various buildings, ranging from small domestic houses to large commercial skyscrapers. Each building has its specific purpose (residential, commercial, or recreational use) and constraints (height limitation reinforced by local government authorities). Thus, the design of each structure can be treated as a unique engineering problem. In nature, structures can be nearly any living organism or products made by them, for example, pine trees or honeybee combs. By studying how these natural structures sustain loads and optimize with resources, structural engineers attempt to innovate existing structural design strategies to achieve efficient and sustainable structures.

For engineers studying biomimicry, three major areas are worth investigating. For example, in organisms, organs and organisms (structures) are made up of different kinds of tissues, which are also made up of cells; a cell is the simplest unit [2]. Structural engineers and builders have used an analogous hierarchy (cell-material, tissue-shape, and organism-structure).

3.1.1. Materials

Materials are the smallest, indistinguishable building blocks in the structure. Natural materials have always been in use. The first tools our ancestors used were little more than sticks or stones picked up and used to hammer open food. In the modern era of engineering, where specific properties are needed for calculations or factors of safety, natural materials have become less popular in structural uses and not obsolete.

Biological materials are elegant and practical in the engineering field. They provide sufficient strength and other special characteristics while remaining relatively light in weight. Most of the natural materials are biodegradable, which increases their value in an era of sustainability. Biomaterials have two main classes: elastic-tensile biomaterials and hard rigid biomaterials. Tensile materials are mainly composed of protein, whereas rigid materials are formed by combining the protein with minerals (primarily calcium or silica) [39].

For example, natural silks have been found to have excellent strength and extensibility [40–42]. Spider silks have low density, tensile strength exceeding 1000 MPa, and extensibility of approximately 0.27, which is way beyond the yield strain of steel (0.0025). In short, natural protein silks are one of the best structural bio-materials made by nature. They have incredible tensile capacity and impressive extensibility, not to mention their low density compared with traditional steel wire. If scientists and engineers can find a way to mass-produce natural silk with a big diameter, the size of concrete reinforcement or cables can be substantially reduced.

Another particular type of protein worth discussing is resilin and abductin, which do not have extremely high tensile strength or elasticity. However, the two types have a special ability to store energy and release it back with high efficiency.

Abalone shell is another great example of nature's wisdom in building construction by using the nearby environment to minimize energy use. Besides its amazing growth mechanism, the abalone has outstanding mechanical properties, as its average fracture strength is 185 MPa [43].

Another type of biomaterials that are also mineralized is bones. Bones are the essential component of our body, performing mechanical, chemical, and biological functions. They are a highly hierarchical structure and have incredible mechanical properties. Bones have two types: cortical (or compact) and cancellous (or trabecular). Bones have a highly hierarchical order. The main components of bones are bone crystals, collagen, and water. The mechanical properties mainly depend on individual bone porosity, degree of mineralization, and bone age [44]. As a result, similar to most biomaterials, their mechanical properties are varied.

Enamel and dentine, (which are known as the stiffest biomaterial in the human body) are mainly found in human teeth, form the other type of bio-materials. Enamel is the stiffest biomaterial in the human body. It has a yield stress of 330 MPa and a Young modulus of 83GPa [45]. Therefore, the yield stress of enamel is comparable to steel. Furthermore, compared with other metals, such as steel, enamel displays metallic-like behavior in a stress-strain relationship and crack initiation even though most enamel is made up of brittle hydroxyapatite crystallites [46]. This finding shows nature's ability to achieve metallic mechanical properties from brittle ingredients using hierarchical order or special arrangement.

These examples show that using natural materials does not necessarily mean a compromise in performance and can indeed be of significant benefit. Thus, why are raw materials not used more today? One major problem of applying the discovery directly to the structural engineering field is that most bio-materials found in nature cannot be mass-produced, and their durability is relatively low. Thus, the problem of mass-producing biomaterials with excellent mechanical properties and increasing their durability will be the most important future research direction for engineers and biomimetics.

The majority of biomimetic materials have been created in Europe, and the form or mechanism of insect or plant organs have been the source of inspiration. Given the rapid advancements in the area of nanotechnology, the biomimetic wave has been also extended to mimicking animals. Japan and the USA are active research participants, and Europe is at the center of growth. The biomimetics research front is proceeded by nanotechnology and dynamically developed using electron microscopes similar to scanning electron microscopy, which allow us to study the physical properties, structure, and function of natural organisms. Given these nanotechnology tools, biomimetic engineers could evaluate using the single-cell scale, especially for organelles of cells and cell interactions. The biomimetic analysis of cell organelles' communities and their structures would provide insights into the development of nanoscale constructs that may act or function during cellular construct performance.

3.1.2. Shape

The shape is the macro arrangement of materials that serve a function. Nature often uses geometrical properties and specific allocation of materials in a macro sense to improve efficiency to resist a combination of loads. For example, plants and animals are constantly under the effect of various loads. The load cases nature often faces are similar to buildings developed by humans. Trees, one of the most stable living structures on the earth, display many structural similarities to load-bearing structures, such as residential buildings. In the case of horizontal branches of trees, the gravity load causes bending stresses along the branches. As the bending axis is relatively stable, horizontal branches develop an elliptical section. The main tree trunk is subjected to wind load as a major lateral load, and its direction is unpredictable. Thus, tree trunks develop circular sections to ensure loading in any direction. The simple mechanics of solids calculations confirm that hollow circular cross-sections are optimal for members subjected to axial, bending, and torsional combined stresses. However, if the direction of loading is known, the hollow elliptical section has an even higher capacity to resist combined stresses, as typically adopted and applied in nature.

Furthermore, to prevent buckling, bamboos develop a hollow section with some nodal septum. Having a nodal septum in equal spacing enhances the buckling resistance of bamboos greatly. It also prevents ovalization from occurring inside the cross-section and stops longitudinal cracks from extending [47].

Besides adopting hollow sections, natural structures often develop tapered members. Cracking is another challenge faced by many organisms. Although no conclusion is drawn about the reason for crack initiation, most scholars have suggested that cracking may result from minor defects or localized damages of structures. In reality, virtually any large structures have cracks. However, cracks allowed to grow or propagate can lead to fracture or reduction in structural capacity. As a result, nature develops several methods to stop crack propagation, such as using composite (laminated) structural materials and placing voids at appropriate places.

Nature has developed excellent strategies to allocate the material to maximize its capacity to resist various mixed actions. Thus, studying how nature combines and arranges biological materials together will provide an understanding of the optimal shape with the optimal proportions of ingredients needed.

3.1.3. Structure

Structural engineers can learn much from nature because it is a self-optimizing system. Nature inspires structural engineers in the process of designing and building a structure that has a high level of adaptability and requires a minimum amount of maintenance. Thus, future structural systems can be considered intelligent, but these basic principles must be first unraveled and understood. Biologists and engineers agree that a systematic analysis in biomimetics is yet to be developed [38].

The structure is the arrangement of different shapes (or members) to solve a given engineering problem. For example, the supporting system of our human body mainly consists of hundreds of bones, ligaments, and tendons. The interrelationship of the bone members has modern applications. The human thigh bone has a high load capacity, which can withstand one ton when in a vertical position. The femur head extends sideways into the hip socket and bears the body's weight off-center. The thigh bone consists of tiny ridges of bone known as trabeculae which are in fact series of studs and braces that are positioned along the force line when standing. The bone structure inspired engineers to decrease the impact of load on the building. In fact, nature strengthens the bone at a level that is required [48].

In 1866, the Swiss engineer Karl Cullman translated these findings into applicable theory. In 1889, French structural engineer Gustave Eiffel was inspired by the concept of "building along the force lines" to design the Eiffel tower. The curve in the Eiffel tower iron is similar to the curve in the femurs' head. The bending and shearing effects caused by the wind would be transformed into compression [48]. From the Eiffel Tower base (Figure 7), the lattice structure of the studs and braces can be seen. The same approach was utilized to design the World Trade Centre.

Bone and joint mechanisms in humans or other animals are one of the most wonderful and simple ways of achieving mobility. It can dislocate the joint under excessive movement or sudden impact without fatal failure in the bone. Furthermore, joints can be relocated again after treatment. Thus, nature has developed a mechanism for repairing and healing itself while adopting simple methods. Suppose this idea can be applied to the field of structural engineering. Thus, engineers may design buildings that can mimic the joint system in humans such that, under the sudden impact (i.e., an earthquake), the building can absorb the energy by dislocating some of its parts and then reconstructing by simply relocating structural connections (joints) back together, which could result in safer buildings and the reduction in construction costs.

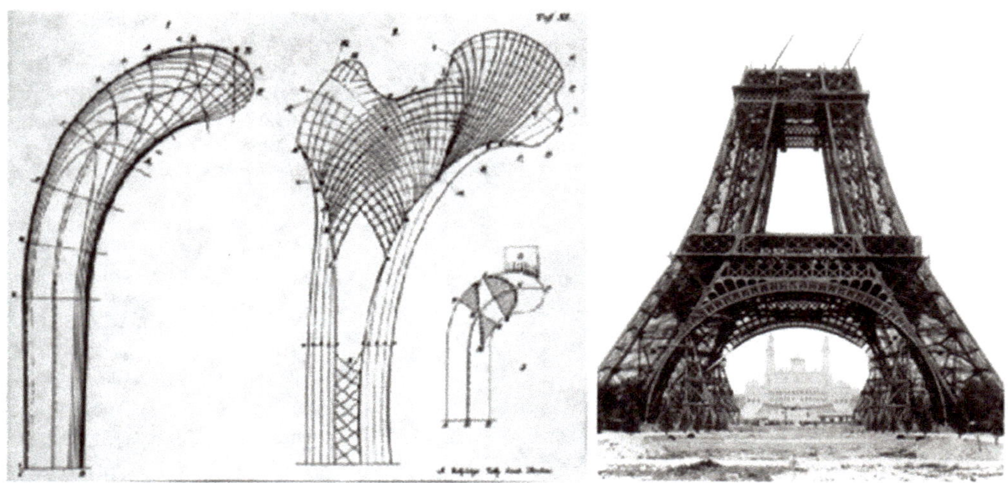

Figure 7. Crane principal stress lines and trabeculae lines on the femur (**left**) [49]; the base of the Eiffel Tower (**right**) (Source: Flickr.com).

Spider webs have some geometric features. Spiders manufacture their webs with two distinctive treads to prevent prey break or bounce back from the web: a strong stiff tread for supporting or structural purposes and the flexible and sticky tread to retain the prey on the web [50].

The tree rooting system is a perfect example of inspiration from nature to withstand loading [36] Compressive buttressing, tensile buttressing, and tap rooting are three mechanisms adopted by various types of trees to resist overturning. Although the structural capacity of the amoeba cell is still under research, it is still a natural wonder that a single-celled organism, without any brain or nervous system, can build a simple yet elegant structure. The strategy of having a spherical shape may also demonstrate the wisdom of nature because a spherical shape is proven to have better resistance on impact from arbitrary directions. If the mechanisms of how amoeba constructs the shell are known, engineers and scientists may develop excellent construction processes, which minimize the need for precise calculation and computation power [28].

Honeycomb is another live structure that inspired many structural engineering projects [51]. A popular application is the utilization of the honeycomb cell for sandwich construction (Figure 8, which is a highly valued structural engineering innovation. The sandwich components are rigidly joined with the core-to-skin adhesive to act as one unit with high rigidity in torsion and bending [52]. Besides saving building material, such a sandwich structure also offers other benefits (i.e., durability, low weight, high stiffness, and stability) compared with usual materials (Figure 8). Thus, materials are used efficiently without sacrificing strength. For instance, bees connect and direct one another through a 'waggle dance' and set up vibrations. Honeycomb is a small dimension structure, and such tremors can be likened to earthquakes. The walls of the honeycomb can absorb these potentially damaging vibrations. This great structure can be imitated when earthquake-proof structures are designed. Jurgen Tautz of the University of Wurzburg in Germany explained that honeybee nest vibrations are similar to low tremors that bees generate. Thus, seeing how the building responds is interesting. Structural engineers can predict building parts that are in danger of earthquakes by considering phase reversal. Consequently, they can strengthen these parts or introduce them into areas that are not critical to absorb damaging vibration [53].

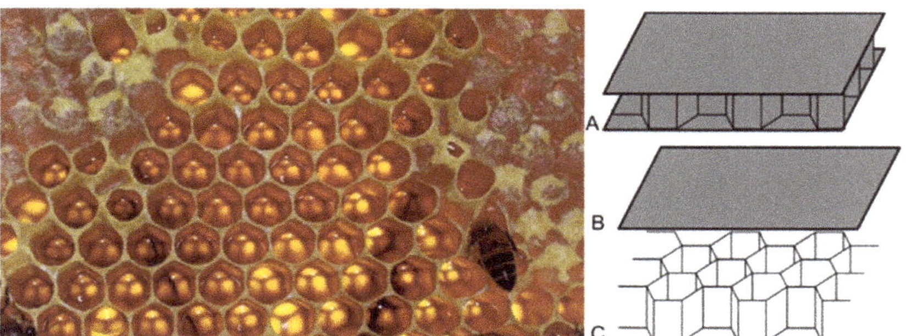

Figure 8. Honeycomb (**left**) and the sandwich structure's schematic view (**right**) (Source: https://en.wikipedia.org/wiki/Sandwich-structured_composite).

4. Biomimicry and Energy Retrofication

Several biomimetic technologies aim to learn from the living world to substitute renewable sources' current fuel energy systems, such as solar and wind energies. Providing energy for buildings and cities is one of the major concerns of today's society mainly because of the urgent need to tackle issues raised by climate change and non-strategic urban planning decisions employed in cities within the last couple of decades.

Biomimetic has also inspired engineers and designers in the development of non-conventional energy resources, such as learning from plant's photosynthesis process to generate solar cell system [54,55], mimicking the butterfly wings in solar panel technology [56], mimicking the ferns in creating more efficient electrodes for solar storage systems [57], mimicking the frog nerve rays in producing batteries, mimicking the foam nests of the Tungara frog and red panda digestive enzymes in the production of biofuels [58], mimicking the movement of fish in creating more efficient wind turbine technologies [59] and mimicking the movement of certain fish in the development of ocean driven energy technologies [60].

Several examples of living ecosystems are highly energy-efficient and are often used as the best inspiration for what humans can do to not depend on fossil fuels. Biomimicry offers great potentials for learning from nature and coming up with solutions that lead to a lower level of energy consumption. Four principles are used in biomimicry to reduce the overall energy consumption; 1—decreasing the demand, 2—identifying unlimited sources of energies; 3—sustainable energy distributing systems, and 4—decreasing the non-toxic flows compatible with w wide range of systems.

Encycle's SwarmLogic utilizes an exceptional algorithm that lets electric appliances interconnect with one another and save power. Almost all major structures have HVAC systems. However, HVAC systems can be the biggest energy consumer and have the highest cost of building maintenance. Various building equipment is operated in isolation from other equipment, following a single timer or thermostat in the facility. Given that these loads do not communicate with one another, they usually operate simultaneously, needlessly increasing energy usage and rising costs. Bee communication in colonies is an inspiration for a building energy management system. Honeybees interconnect and manage individual behaviors to shape a collective organization that effectively feeds colonies and builds hives.

The project developed Swarm Logic to lessen the demand for peak energy utilization by up to 30%. The energy-efficient technology integrates a structure's controls to instantly and dramatically reduce power costs. Controllers of Swarm Logic establishes a wireless mesh network of electric-consuming appliances and enables intercommunication autonomously. The interconnected appliances spread out power demand through a custom algorithm that is inspired by honeybee communication. The outcome is referred to as peak demand shaving.

One of the most well-known examples of biomimicry in reducing energy consumption in buildings is the CH2 building in Melbourne, a six green star building, which was built at a cost premium of 22.1%. However, given that productivity increases of 10.9% from staff attributed to the new building, payback was between 5 and 7 years [61,62]. In this building, the air is conditioned through the use of cleaned water in the sewage system. This process has been inspired by certain termite species that employ aquifer water as an evaporative cooling mechanism. Termite digs a deep tunnel to reach the water and therefore, its cooling impact is a remedy to reduce the extreme heat and keep the mound within a one-degree temperature variation range.

The concept of Biomimicry has also inspired the use of renewable energy systems in the built environment and technologies and led to significant savings in energy usage. For example, the tubercles on the flippers of humpback whales have been the main inspiration for a type of wind turbine in the project of reference [63]. Most wind turbines stop working under low wind scenarios. However, this project's wind turbine blade has been designed so that the performance is not adversely affected even under slower speeds. This project achieved a 20% improvement in the annual output due to employing biomimicry-inspired principles in design and construction.

The concept of "Green Power Island" is based on the necessity to provide diverse forms of energy storage to accommodate a different range of renewable energy outputs and generate resilient systems. The proposal provides a solution to this challenge by creating a large reservoir with 22,000,000 m^3 capacity and a generation capacity of 2.3 GWh. The reservoir generates power by letting the sea flood back in via turbines that are located in the flatlands around the reservoir and provide the best access to the wind. The site next to the turbines is also used as a platform to grow biomass and food crops and thereby deliver multiple benefits. The reservoir is also equipped with series of photovoltaics that enables the possibility of solar tracking. Surrounding lands of the island provides the best platform to breed seabirds. Below the sea level, the sloping border walls have created a rocky shoreline. There is a lack of biodiversity in flat rocky seabeds, and the rocky shorelines are in contrast known as rich habitats. Therefore, this proposal can effectively enhance biodiversity [1].

5. Summary and Conclusions

This study presented biomimicry's potential to provide sustainable solutions to human challenges, especially designing and constructing structures. Biomimicry is also helpful in the creation of novel materials, technologies, and products with viable attributes. However, biomimicry knowledge is lacking among stakeholders in architecture and structural engineering. Thus, biomimicry's adoption and application are hindered from enhancing sustainability in the construction and design industries. Principles of biomimicry also play an essential role in evaluating sustainability because they are common tools and vital checklists that are strictly used when focusing on sustainability. Professional awareness, education, and training on biomimicry of stakeholders and professionals should be stimulated to ensure its wide adoption and practice.

This review showed that biomimicry's connectivity has been supported throughout history. After the Renaissance, humans developed a better understanding of physics and mathematics. They were able to form large metal products with fuel-powered machinery with the aid of the industrial revolution [38]. As a result, structural engineers and designers have preferred working with forms, shapes, and materials with uniform properties throughout and members with easily determined mechanical properties. Principles in nature's prototypes were often excessively complex to be transferred for engineering and design purposes; built environment professionals often have members with homogenous non-composite materials, shapes and forms with rigid rectangular shapes as opposed to using composite, flexible, and force adaptive members, commonly found in nature [64].

Nature builds things more gently. Biomimetics has discovered that nature recycles everything used, uses sunlight as the primary energy resource, and fits its function [2].

Biomimicry also aims to provide innovative, sustainable solutions to engineering problems by studying biological modes and systems found in nature. Therefore, biomimicry has a great potential to benefit structural engineering and the design process.

Architects and structural engineers require great awareness to achieve efficiency and sustainability in buildings, especially in this era when meeting the sustainability targets is more critical than in the past. The natural world provides an extensive design database that can inspire creative thoughts. Most efficient buildings adapt to their surrounding environment and use the environment to benefit them. Consequently, the value of the lessons from nature should be recognized, and these innovations should be adopted in developing structures that fit with their surroundings.

It is also worthwhile mentioning that the majority of functional mimicries were derived from insects and plants' micro-and nanoscale parts. Recently, given the nanotechnology advancement, a new wave of biomimetics has been extended to animal imitation. Europe has been the center of development, especially of most biomimetic materials. Japan and the USA actively participate in research. Biomimetic research front is vigorously advanced by nanotechnology and electron microscopes similar to scanning electron microscopy, enabling the observation and analysis of natural organisms' physical properties, structure, and function. These tools of nanotechnology allow biomimetic engineers to study at the single-cell scale for cell organelles and interactions. The analysis of cell organelle communities and their structures offers us insights into developing nanoscale constructs during the performance of cellular constructs.

The design and management of future cities could also incorporate biomimicry but may have big obstructions. Adopting transdisciplinary solutions needs considerable changes to city powers and the cooperation among stakeholders, systems, and utility providers. Furthermore, the public, local authorities, developers, and designers must have adaptive mindsets to exploit biomimicry's potential fully.

Author Contributions: E.J. and Z.V. conceptualizing and critically reviewing the literature, writing the first draft of the manuscript, technical revision and editing the manuscript. All authors have read and agreed to the published version of the manuscript.

Funding: This research received no external funding.

Institutional Review Board Statement: Not applicable.

Informed Consent Statement: Not applicable.

Data Availability Statement: Not applicable.

Conflicts of Interest: The authors declare no conflict of interest.

References

1. Pawlyn, M. *Biomimicry in Architecture*; Routledge: Oxfordshire, UK, 2019.
2. Benyus, J.M. *Biomimicry: Innovation Inspired by Nature*; Morrow: New York, NY, USA, 1997.
3. Klein, L. *A Phenomenological Interpretation of Biomimicry and Its Potential Value for Sustainable Design*; Kansas State University: Manhattan, Kansas, 2009.
4. El-Zeiny, R.M.A. Biomimicry as a problem solving methodology in interior architecture. *Procedia-Soc. Behav. Sci.* **2012**, *50*, 502–512. [CrossRef]
5. Kennedy, E.; Fecheyr-Lippens, D.; Hsiung, B.-K.; Niewiarowski, P.H.; Kolodziej, M. Biomimicry: A path to sustainable innovation. *Des. Issues* **2015**, *31*, 66–73. [CrossRef]
6. Gruber, P. *Biomimetics in Architecture*; Ambra Verlag: Barcelona, Spain, 2010.
7. Vierra, S. Biomimicry: Designing to model nature. *Whole Build. Des. Guide* **2011**, 1–10.
8. Nachtigall, W. Bionik—Was ist das? In *Bionik*; Springer: New York, NY, USA, 1998; pp. 3–15.
9. Murr, L. Biomimetics and biologically inspired materials. In *Handbook of Materials Structures, Properties, Processing and Performance*; Springer: Cham, Switzerland, 2015; pp. 521–552.
10. Nachtigall, W. *Grundlagen und Beispiele für Ingenieure und Naturwissenschaftler*; Springer: New York, NY, USA, 2002.
11. Zari, M.P. Can biomimicry be a useful tool for design for climate change adaptation and mitigation? In *Biotechnologies and Biomimetics for Civil Engineering*; Springer: New York, NY, USA, 2015; pp. 81–113.
12. Otto, F. *Occupying and Connecting*; Edition Axel Menges: Fellbach, Germany, 2003.

13. Feuerstein, G. *Biomorphic Architecture*; Edition Axel Menges: Fellbach, Germany, 2002.
14. Badarnah Kadri, L. *Towards the LIVING Envelope: Biomimetics for Building Envelope Adaptation*; Citeseer: Princeton, NJ, USA, 2012.
15. Zari, M.P. Biomimetic approaches to architectural design for increased sustainability. In Proceedings of the SB07 NZ Sustainable Building Conference, Auckland, New Zealand, 14–16 November 2007; pp. 1–10.
16. Knippers, J. Building and Construction as a Potential field for the Application of Modern Bio mimetic Principles. In *International Biona Symposium*; University of Stuttgart: Stuttgart, Germany, 2009.
17. Helms, M.; Vattam, S.S.; Goel, A.K. Biologically inspired design: Process and products. *Des. Stud.* **2009**, *30*, 606–622. [CrossRef]
18. Casey, V. *Biomimicry 3.8: What Would You Ask Nature*; Core, 2012. Available online: https://www.core77.com/posts/21799/biomimicry-38-what-would-you-ask-nature-21799 (accessed on 9 August 2021).
19. Faludi, J. Biomimicry for green design (a how to). *World Chang.* **2005**, *200*.
20. Panchuk, N. *An Exploration into Biomimicry and Its Application in Digital & Parametric [Architectural] Design*; University of Waterloo: Waterloo, ON, Canada, 2006.
21. Zari, M.P. *Regenerative Urban Design and Ecosystem Biomimicry*; Routledge: Oxfordshire, UK, 2018.
22. Hensel, M.; Sunguroglu, D.; Menges, A. Material Performance. *Archit. Des.* **2008**, *78*, 34–41. [CrossRef]
23. Persiani, S. *Biomimetics of Motion: Nature-Inspired Parameters and Schemes for Kinetic Design*; Springer: New York, NY, USA, 2018.
24. Aldersey-Williams, H. Towards biomimetic architecture. *Nat. Mater.* **2004**, *3*, 277–279. [CrossRef]
25. Pohl, G.; Nachtigall, W. *Biomimetics for Architecture & Design: Nature-Analogies-Technology*; Springer: New York, NY, USA, 2015.
26. Naboni, R.; Paoletti, I. *Advanced Customization in Architectural Design and Construction*; Springer: New York, NY, USA, 2015.
27. Lang, N.; Pereira, M.J.; Lee, Y.; Friehs, I.; Vasilyev, N.V.; Feins, E.N.; Ablasser, K.; O'Cearbhaill, E.D.; Xu, C.; Fabozzo, A. A blood-resistant surgical glue for minimally invasive repair of vessels and heart defects. *Sci. Transl. Med.* **2014**, *6*, 218ra216. [CrossRef] [PubMed]
28. Hansell, M.; Hansell, M.H. *Animal Architecture*; Oxford University Press on Demand: Oxford, UK, 2005.
29. Garfield, C.A. *Peak Performers: The new Heroes of AMERICAN Business*; William Morrow & Company: New York, NY, USA, 1986.
30. Sorguç, A.G.; Hagiwara, I.; Selcuk, S. Origamics in architecture: A medium of inquiry for design in architecture. *Metu Jfa* **2009**, *2*, 235–247. [CrossRef]
31. Gruber, P.; Jeronimidis, G. Has biomimetics arrived in architecture? *Bioinspir. Biomim.* **2012**, *7*, 010201. [CrossRef]
32. Rodin, J. The Resilience Dividend: Managing Disruption. *Avoid. Disasterand* **2014**.
33. Kindle, E.M. Notes on the point Hope spit, Alaska. *J. Geol.* **1909**, *17*, 178–189. [CrossRef]
34. Boyce, P.R. The impact of light in buildings on human health. *Indoor Built Environ.* **2010**, *19*, 8–20. [CrossRef]
35. Parker, A.R.; Lawrence, C.R. Water capture by a desert beetle. *Nature* **2001**, *414*, 33–34. [CrossRef]
36. Vogel, S. *Comparative Biomechanics: Life's Physical World*; Princeton University Press: Princeton, NJ, USA, 2013.
37. Oguntona, O.A.; Aigbavboa, C.O. Biomimicry principles as evaluation criteria of sustainability in the construction industry. *Energy Procedia* **2017**, *142*, 2491–2497. [CrossRef]
38. Yiatros, S.; Wadee, M.A.; Hunt, G.R. The load-bearing duct: Biomimicry in structural design. In *Proceedings of the Institution of Civil Engineers-Engineering Sustainability*; Thomas Telford Ltd: London, UK, 2007; pp. 179–188.
39. Al-Ketan, O.; Rowshan, R.; Alami, A.H. *Biomimetic Materials for Engineering Applications*; Elsevier: Amsterdam, The Netherlands, 2020.
40. Xu, M.; Lewis, R.V. Structure of a protein superfiber: Spider dragline silk. *Proc. Natl. Acad. Sci. USA* **1990**, *87*, 7120–7124. [CrossRef] [PubMed]
41. Hakimi, O.; Knight, D.P.; Vollrath, F.; Vadgama, P. Spider and mulberry silkworm silks as compatible biomaterials. *Compos. Part B Eng.* **2007**, *38*, 324–337. [CrossRef]
42. Gould, J.L.; Gould, C.G. *Animal Architects: Building and the Evolution of Intelligence*; Basic Books: New York, NY, USA, 2012.
43. Meyers, M.A.; Chen, P.-Y.; Lin, A.Y.-M.; Seki, Y. Biological materials: Structure and mechanical properties. *Prog. Mater. Sci.* **2008**, *53*, 1–206. [CrossRef]
44. Currey, J. Incompatible mechanical properties in compact bone. *J. Theor. Biol.* **2004**, *231*, 569–580. [CrossRef]
45. Staines, M.; Robinson, W.; Hood, J. Spherical indentation of tooth enamel. *J. Mater. Sci.* **1981**, *16*, 2551–2556. [CrossRef]
46. He, L.H.; Swain, M.V. Enamel—A "metallic-like" deformable biocomposite. *J. Dent.* **2007**, *35*, 431–437. [CrossRef] [PubMed]
47. Schulgasser, K.; Witztum, A. On the strength, stiffness and stability of tubular plant stems and leaves. *J. Theor. Biol.* **1992**, *155*, 497–515. [CrossRef]
48. Meadows, R. Designs from life. *Zoogoer* **1999**, *28*, 286–289.
49. Skedros, J.G.; Baucom, S.L. Mathematical analysis of trabecular 'trajectories' in apparent trajectorial structures: The unfortunate historical emphasis on the human proximal femur. *J. Theor. Biol.* **2007**, *244*, 15–45. [CrossRef] [PubMed]
50. Zschokke, S. Form and function of the orb-web. *Eur. Arachnol.* **2000**, *19*, 99.
51. Peterson, I. The honeycomb conjecture: Proving mathematically that honeybee constructors are on the right track. *Sci. News* **1999**, *156*, 60–61. [CrossRef]
52. Hexcel. Honeycomb Sandwich Design Technology. Publication No. AGU 075b, 2000.
53. Klarreigh, E. *Good Vibrations, Nature Science Update*; Nature Science: London, UK, 2001.
54. Llansola-Portoles, M.J.; Gust, D.; Moore, T.A.; Moore, A.L. Artificial photosynthetic antennas and reaction centers. *C. R. Chim.* **2017**, *20*, 296–313. [CrossRef]

55. Martín-Palma, R.J.; Lakhtakia, A. Engineered biomimicry for harvesting solar energy: A bird's eye view. *Int. J. Smart Nano Mater.* **2013**, *4*, 83–90. [CrossRef]
56. Shanks, K.; Senthilarasu, S.; Mallick, T.K. White butterflies as solar photovoltaic concentrators. *Sci. Rep.* **2015**, *5*, 1–10. [CrossRef]
57. Thekkekara, L.V.; Gu, M. Bioinspired fractal electrodes for solar energy storages. *Sci. Rep.* **2017**, *7*, 1–9. [CrossRef]
58. Wendell, D.; Todd, J.; Montemagno, C. Artificial photosynthesis in ranaspumin-2 based foam. *Nano Lett.* **2010**, *10*, 3231–3236. [CrossRef] [PubMed]
59. Whittlesey, R.W.; Liska, S.; Dabiri, J.O. Fish schooling as a basis for vertical axis wind turbine farm design. *Bioinspir. Biomim.* **2010**, *5*, 035005. [CrossRef]
60. Allen, M.J. *Continental shelf and upper slope. The Ecology of Marine Fishes: California and Adjacent Waters*; University of California Press: Berkeley, CA, USA, 2006; pp. 167–202.
61. Aranda-Mena, G.; Crawford, J.; Chevez, A.; Froese, T. Building information modelling demystified: Does it make business sense to adopt BIM? *Int. J. Manag. Proj. Bus.* **2009**, *2*, 419–434. [CrossRef]
62. Paevere, P.; Brown, S. Indoor environment quality and occupant productivity in the CH2 building: Post-occupancy summary. In Proceedings of the 2008 International Scientific Committee World Sustainable Building Conference, Melbourne, Australia, 21–25 September 2008.
63. Ju, J.; Bai, H.; Zheng, Y.; Zhao, T.; Fang, R.; Jiang, L. A multi-structural and multi-functional integrated fog collection system in cactus. *Nat. Commun.* **2012**, *3*, 1–6. [CrossRef] [PubMed]
64. Milwich, M.; Speck, T.; Speck, O.; Stegmaier, T.; Planck, H. Biomimetics and technical textiles: Solving engineering problems with the help of nature's wisdom. *Am. J. Bot.* **2006**, *93*, 1455–1465. [CrossRef]

Article

Daylighting Performance of Light Shelf Photovoltaics (LSPV) for Office Buildings in Hot Desert-Like Regions

Abdelhakim Mesloub [1],* and Aritra Ghosh [2],*

[1]   Department of Architectural Engineering, Ha'il University, Ha'il 2440, Saudi Arabia
[2]   College of Engineering, Mathematics and Physical Sciences, Renewable Energy, University of Exeter, Penryn, Cornwall TR10 9FE, UK
*   Correspondence: a.maslub@uoh.edu.sa (A.M.); a.ghosh@exeter.ac.uk (A.G.)

Received: 19 October 2020; Accepted: 5 November 2020; Published: 10 November 2020

Abstract: Visual comfort and energy consumption for lighting in large office buildings is an area of ongoing research, specifically focusing on the development of a daylight control technique (light shelf) combined with solar energy. This study aims to investigate the optimum performance of light shelf photovoltaics (LSPV) to improve daylight distribution and maximize energy savings for the hot desert-like climate of Saudi Arabia. A radiance simulation analysis was conducted in four phases to evaluate: appropriate height, reflector, internal curved light shelf (LS) angle, and the integrated photovoltaic (PV) with various coverages (25%, 50%, 75%, and entirely external LS). The results revealed that the optimum is achieved at a height of 1.3 m, the addition of a 30 cm reflector on the top of a window with an internal LS curved angle of 10° with 100% coverage (LSPV1, LSPV2). Such an arrangement reduces the energy consumption by more than 85%, eliminates uncomfortable glare, and provides uniform daylight except for during the winter season. Hence, the optimization of the LSPV system is considered to be an effective solution for sustainable buildings.

Keywords: LSPV; visual comfort; energy consumption; angle

1. Introduction

Presently, the Kingdom of Saudi Arabia is witnessing unprecedented development as part of the implementation of its ambitious 2030 vision. Launched in 2016, the vision intends to build a thriving, diversified, and sustainable economic model which has a lesser dependence on the income from oil and savings from fossil fuel subsidies [1]. Solar energy has been recognized as a relatively significant source of renewable energy compared to the other sustainable energy sources in the Kingdom of Saudi Arabia. Considering the ample solar radiance and the prevailing desert-like climatic conditions [2], photovoltaic (PV) technology could play a vital role in overcoming energy issues in this hot and arid region [3]. The Saudi Electricity Company revealed that lighting accounts for over 30% of the total energy consumed in office buildings to provide the required amount of light [4]. Therefore, the rationalization of energy use and the improvement of daylighting design techniques such as light shelves [5–7], semi-transparent photovoltaics (STPV) [8], louvers [9], and light transportation [10] need to be considered.

One of the essential daylighting techniques is light shelves. A light shelf is a horizontal surface that reflects daylight deep inside a building. Light shelves are placed above eye-level and have highly reflective upper surfaces that reflect daylight onto the ceiling and deeper into a space. It is a flexible system that offers a variety of design solutions. It can be installed easily on an external or internal surface [6], and can be designed in various geometries like fixed flat models and curved reflective surfaces [11], dynamic shading device design [5,12], and integrated as a solar module [13]. Thus, precise operation of the various design solutions associated with light shelves, such as solar modules, is required to enhance lighting and electricity generation efficiency in response to external conditions, as depicted in Figure 1.

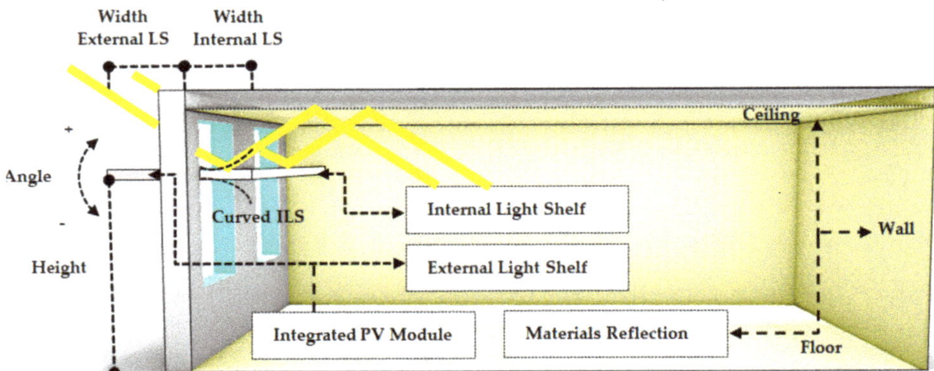

Figure 1. Design solutions and variables associated with a light shelf that influence the daylight distribution.

Much research has been conducted concerning the daylight performance of light shelves using several configurations (dimensions, geometry, angles, and material) using experiments in the field under varying sky conditions or through simulations. As specified in Table 1, previous research has primarily evaluated the optimization of the aspects concerning light shelves to enhance the distribution of daylight in vast spaces to facilitate enhanced visual comfort. Nevertheless, there was specific research concerning a light shelf whose width and length was modified by isolating the light shelf reflector along with the bottom and top reflectors [14].

Several other studies analyzed an integration of building envelope technologies like louvres [15], in-built photovoltaics (PV) [13], and awning systems [16] to facilitate a decrease in the energy requirements of the building. A review of the existing literature indicates that there are only a few pieces of research that assess the performance of solar panels integrated with light shelves. Hwang et al. [17] examined the characteristics of the light shelf system using in-built PVs by lining the outer light shelf with solar panels in a southern orientation, placed at varying angles. Heangwoo Lee [13] recently suggested a solar-integrated light shelf having only a fraction of it covered with solar panels at varying angles; however, the internal shelf would not be used. Hence, the current study aims to assess the optimum performance of the combination of external and internal light shelves using integrated solar panels in several places to facilitate the appropriate distribution of daylight in the office structure and optimize energy savings, especially under the hot desert-like climatic conditions of Ha'il city.

Table 1. Studies conducted on light shelf performance on daylighting.

Title of Study and Year	Light Shelf Parameters	Experimental Measurement	Daylighting Simulation	Integrated PV with Light Shelf
Maximizing the light shelf performance by interaction between light shelf geometries and a curved ceiling, Jordan, 2010 [11]	Geometries, curved ceiling	no	yes	no
Power performance of photovoltaic-integrated lightshelf systems, South Korea, 2014, [17]	Full solar panel (one size)	yes	yes	yes
Analysis of the Impacts of Light Shelves on the Useful Daylight Illuminance in Office Buildings in Toronto, 2015, [18]	Width, depth, and height	no	yes	no

Table 1. *Cont.*

Title of Study and Year	Light Shelf Parameters	Experimental Measurement	Daylighting Simulation	Integrated PV with Light Shelf
Dynamic internal light shelf for tropical daylighting in high-rise office buildings, Malaysia, 2016, [5]	Height of ILS, number of ILS	yes	yes	no
Evaluating daylight performance of light shelves combined with external blinds in south-facing classrooms in **Athens**, 2016, [15]	Material, angle, combination with louvers	no	yes	no
Investigating the Influence of Light Shelf Geometry Parameters on Daylight Performance and Visual Comfort, a Case Study of Educational Space in Tehran, **Iran**, 2016, [19]	Dimension, angle, orientation	no	yes	no
Energy-saving performance of light shelves under the application of user-awareness technology and light-dimming control, South Korea, 2018, [20].	Angle	yes	yes	no
A preliminary study on the performance of an awning system with a built-in light shelf, South Korea, 2018, [16].	Angle, awning system	yes	yes	no
Development and Performance Evaluation of Light Shelves Using Width-Adjustable Reflectors, South Korea, 2018, [21].	Width adjustability	yes	no	no
Performance evaluation of a light shelf with a solar module based on the solar module attachment area, South Korea, 2019, [13].	Angle, PV coverage	yes	yes	yes
Experimental Analysis of the Performance of Light Shelves in Different Geometrical Configurations Through the Scale Model Approach, South Korea, 2020, [6].	Angle, curved LS, ILS reflector	yes	no	no
Optimization of Daylight Performance Based on Controllable Light-shelf Parameters using Genetic Algorithms in the Tropical Climate of **Malaysia**, 2020, [22]	Angle, width/depth ratio, height, and reflectivity	yes	yes	no
Performance Evaluation of External Light Shelves by Applying a Prism Sheet, South Korea, 2020, [23].	Applying prism sheets with angles	yes	no	no

1.1. The Sun Path of Ha'il City

Sun path diagrams are appropriate for depicting the annual changes happening about the Sun's path. As shown in Figure 2, the midday sun in Ha'il city is at a 39° inclination during the winter solstice. The inclination is 84° during the summer solstice. These numbers indicate a 45° deviation between the solstices as a result of the Earth's declination, varying between −23° and +23° over the year. In contrast, during spring and autumn, the Sun has a similar path and an inclination of 63°, which is present between the two solstices.

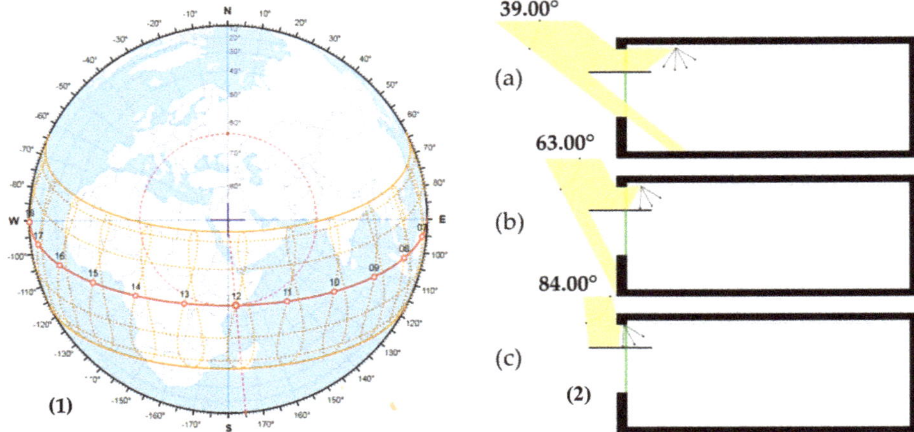

Figure 2. (**1**) Sun path at the city of Ha'il using Andrew–Marsh 2D modeling, (**2**) the inflow of daylight corresponding to the incident angle of design rays at midday. (**a**) Winter solstice; (**b**) spring/autumn equinox; (**c**) summer solstice.

The solar energy incident on the southern and northern parts of the building is not the same since the southern side has better incident sunlight compared to the northern side. In contrast, considering the eastern and western directions, the incident solar energy exhibits symmetry. As a result, there should be a difference in the light intensity once the light shelves are attached.

1.2. Recommended Indoor Illuminance

According to the International Organization for Standardization (ISO) standard 8995-1:2002 [24], typically, the illumination at the workplace (especially in areas comprising continuous work) should not fall below 300 lux; however, the delta between the suggested illuminances is not substantial because they are usually associated with the Commission International d'Eclairage (CIE) regulations. Recommended illumination in the United States [25], Japan [26], and European countries [27] range between 300 lux to 1000 lux. For most countries whose data were reviewed, 500 lux illumination is maintained at the desk level for routine office work, as presented in Appendix A. Most countries recommend 750 lux for drawing since it is a task that requires high accuracy. In contrast, the lower limit of horizontal illumination in an office setting concerning a computer lies in the 300–750 lux range. In this study, the indoor illumination standard for light shelf performance evaluation was set to between 300–750 lux based on the review of the aforementioned optimum indoor illumination standards.

2. Materials and Methods

This research considers both qualitative and quantitative analyses of the simulation of numerical parameters (reflector, height, and Internal light shelf (ILS) curve) as a mechanism for three-phased data collection. Subsequently, the PV modules are integrated at varying coverage values.

The following sections present the simulation technique, the light shelf configurations used for the case study, and the assessment criteria implemented in this study. Figure 3 provides a summary of the chosen case.

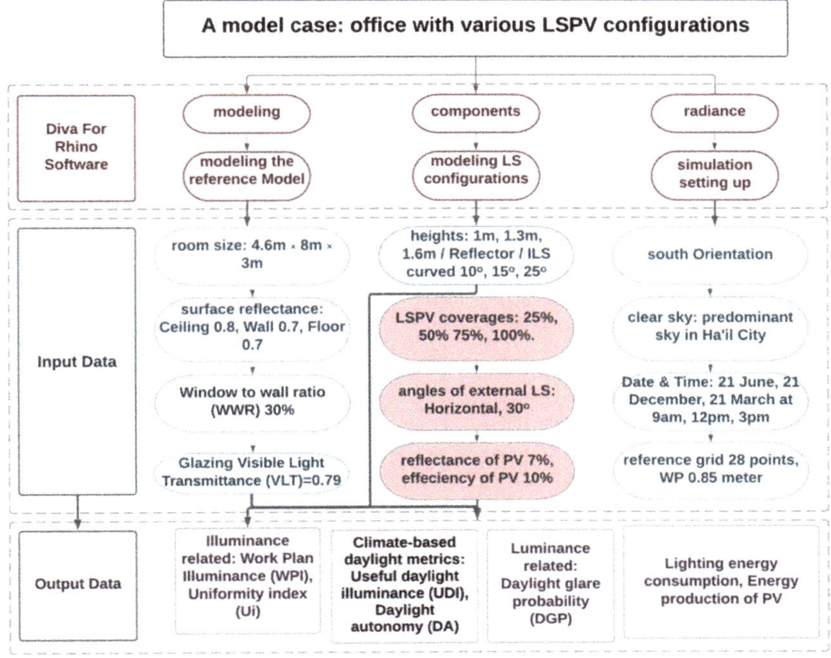

Figure 3. Schematic diagram for the simulation method.

2.1. Computer Simulation (Diva for Rhino)

This research used Diva for Rhino as the tool to simulate daylight performance for different light shelf configurations. Diva employs a Radiance backwards ray-tracer calculation method and the DAYSIM daylighting analysis engine to run daylight simulation [28]. The software considers the distribution of emitted rays and supports the analysis of reflection, transmission and refraction from surfaces. The Meteonorm metrological database (TMY) is used by Diva, where the hourly data of weather aspects like radiance, illuminance, temperature and humidity are validated by the rating authority [29].

2.2. Light Shelf Configurations

In this study, 11 internal–external light shelf configurations were tested using daylight simulation, as shown in Table 2. The combined internal and external light shelf (LS) configurations have been assessed using four main phases:

Table 2. Light shelf configurations used in simulation for each phase.

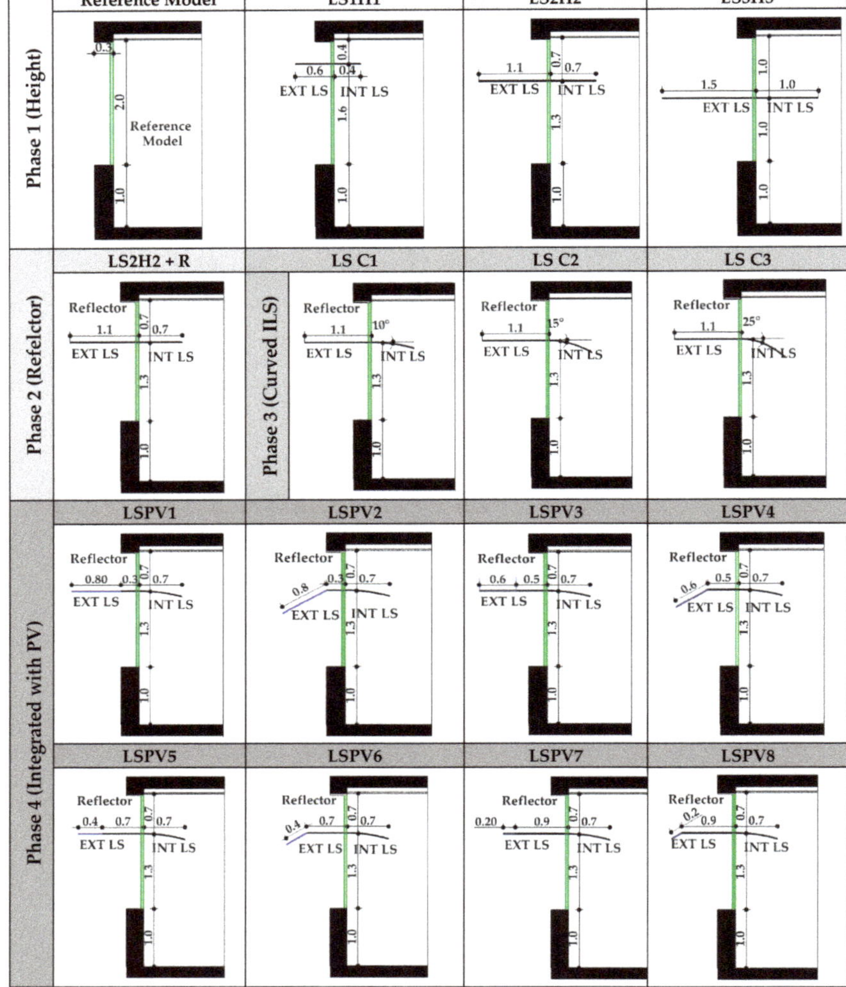

- First phase: three horizontal light shelves of different dimensions, placed at varying heights as defined by previous studies [30–32]. The Internal LS depth is equal to the height of the clerestory window above it while the external LS depth is 1.5 times the height of the clerestory window.

$$d_{\text{int light shelf}} = h_{\text{clerestory}}$$

$$d_{\text{ext light shelf, max}} \leq 1.5 \times h_{\text{clerestory}}$$

The three configurations in the first phase are LS1H1, LS2H2, and LS3H3, and they have fixed heights of 1.6 m, 1.3 m, and 1.0 m, respectively.

- Second Phase: once the appropriate height of the LS is chosen, a reflector material is added on the top of the window to enhance the illumination of the daylit area at the back and then compared with the reference model.
- Third phase: three downward-curved internal LS (having angles of 10°, 15° and 25°) were examined and compared with horizontal internal LSs with the reflector material specified in the second phase.
- Fourth Phase: four different PV coverage and tilt angles are integrated with the external LS. LSPV1, LSPV3, LSPV5, and LSPV7 are designed to have external horizontal light shelves with 100%, 75%, 50%, and 25% coverage, respectively. The remaining configurations of LSPVs have the same coverage but are tilted at 30° to reflect maximum energy and prevent the occurrence of glare.

2.3. Description of the Case Study and Climatic Conditions

The model chosen for reference in this case study considers a typical vast office structure built in 'Diva for Rhino' with a depth of 8.0 m and a width of 4.6 m. As depicted in Figure 4, the office was aligned with the southern façade using a lateral typology. In the context of the reference model, the window-to-wall ratio (WWR) is 0.3. At the same time, the opening is double glazed using low-E material, and has a visible light transmittance (VLT) value of 0.79. Table 3 lists the radiance parameters used for daylighting simulation.

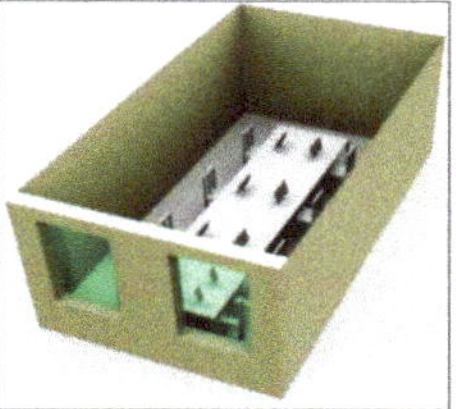

Figure 4. Exterior and interior 3D reference model and furniture arrangement.

Table 3. Radiance parameters used in the daylighting simulation.

Radiance Parameter	Ambient Bounces	Ambient Divisions	Ambient Sampling	Ambient Accuracy	Ambient Resolution
Value	7	1500	100	0.1	300

This study used data pertaining to the Ha'il area as representative of the hot desert-like climate of Saudi Arabia, which is summarized in Table 4. This area is located around the center of the Arabian Peninsula, and its coordinates are 27°31′ N, 41°41′ E. The average temperature is 31.1 °C and the coldest month is January, which has a mean temperature of 10.6 °C. The area receives intense solar radiation—the average monthly incident solar energy experiences significant seasonal variation over the course of the year. The monthly maximum global horizontal radiation values range from 243 kWh/m^2 in summer and 118 kWh/m^2 in winter. At the same time, the global diffuse radiance is approximately 29%. There is shallow cloud cover in this area. The area experiences clear skies on more than 70% of the days in an average year, while 29% of the days are overcast or mostly overcast; hence, this area has clear sky conditions according to the CIE standard [33]. Furthermore, the daylight range is between 10.4–13.6 h, which indicates plenty of daylight through the year.

Table 4. Monthly climatic conditions of Ha'il city.

	January	February	March	April	May	June	July	August	September	October	November	December
Average air temperature (°C)	10.2	13.5	17.8	23.3	28.1	31.9	33.2	33.8	30.6	25.1	17	12.2
Global horizontal irradiance (kWh/m^2)	125	132	183	204	228	243	238	225	195	168	125	118
Global diffuse irradiance (kWh/m^2)	38	40	53	62	64	56	61	60	51	45	40	29
Sun hours	10.6	11.2	12	12.8	13.5	13.8	13.7	13.1	12.3	11.5	10.8	10.4
Cloud cover (%)	26	22	23	24	17	3	10	9	6	17	26	29

2.4. Modelling Approach and Analysis Criteria

The simulated work plane illuminance (WPI) values were transcribed into a tabular form which contains the average illumination of four grid points in each row (lines as shown in Figure 5a). Furthermore, the quality of daylight distribution employed the uniformity index (Ui) of interior daylight, which is the ratio of the average illumination to the maximum illumination. This ratio should not be less than 0.6 above the work plane, as per the NBN L13-001 code and international guidelines [34]. The simulations were performed within the design days at the summer and winter solstices and during the mid-season (21 June, 21 December, 21 March) at 9.00 a.m., 12.00 p.m., and 3.00 p.m. to evaluate the annual variation and the critical period. The choice of the simulated points (grids) inside the office followed the grids plotted as per the arrangement of work planes. The minimum number of points in the deep-office prototype was 28. The distance between the simulated grid points was kept at 1 m for accurate results, as depicted in Figure 5. The reflection by the surfaces used in this experimental set is presented in Table 5.

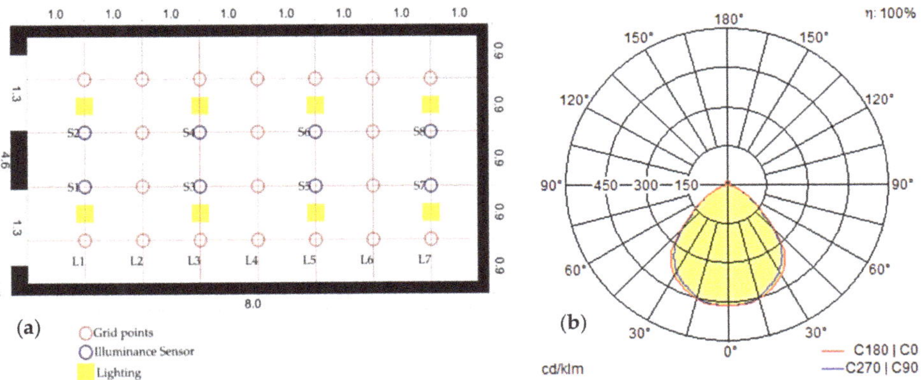

Figure 5. (**a**) Plan view representation of the grid points, illuminance sensors and lighting positions, (**b**) light distribution of conical illuminance.

Table 5. Material reflection coefficient percentages.

Material	Reflection Coefficient
Ceiling	80%
Floor	20%
Wall	70%
Furniture	50%
Light shelf	90%
Solar panel	07%
Reflector	99%
Double glazing low-E	VLT 79%

For more details, the analysis of LSPV in last phase used climate-based daylight metrics that include the useful daylight illuminance (UDI), daylight autonomy (DA) and daylight glare probability (DGP), supported by a 3D illuminance contour map for glare assessments. The criteria of assessments of each metric are summarized in Table 6. The simulation of lighting energy consumption is through daylight control strategy (photosensor controlled dimming: 300 lux), where eight light-emitting diode (LED) indoor artificial lights (17.5 W luminaire power, luminaire luminous flux: 1552 lm and luminaire efficacy: 89 lm/W) were placed in the middle of line 1, 3, 5 and 7 and linked with illumination sensors (S1, S2, S3, S4, S5, S6, S7 and S8). The light distribution of the lighting in this study is shown in

Figure 5b. The illumination sensors were placed 0.85 m above the floor surface, based on the height of the work plane. Their operating profiles follow their real use from 8.00 a.m. to 5.00 p.m.

Table 6. The performance indicators of visual comfort used in this study.

Criteria	Performance Indicator of Delighting Quantity and Quality
WPI	WPI recommended 300–750 lux
UI	Uniformity index must be greater than 0.5
UDI	100 lux < dark area (need artificial light) 100 lux–2000 lux (comfortable), at least 50% of the time >2000 lux too bright with thermal discomfort
DA	Set up 300 lx
DGP	0.35 < imperceptible glare 0.35–0.40 perceptible glare 0.4–0.45 disturbing glare >0.45 intolerable glare
Net energy	Energy production of solar panels—energy consumption of artificial lighting

The amount of energy produced by the multi-crystalline photovoltaic module was derived using the Simple model. This model was simulated by multiplying the fraction of surface area that had active solar cells by the total solar radiation incident on the PV array; module conversion efficiency was set to $n = 10\%$. While the area of the solar module used varies according to the case set used in this study, the amount of energy produced depends primarily on photovoltaic array efficiency and inverter modelling efficiencies at the operating conditions. Nevertheless, the difference between lighting energy consumption and the energy produced is either positive or negative. Hence, for every configuration, the net energy production is the most significant performance indicator required to determine the optimum LSPV configuration.

3. Results and Discussion

3.1. Analysis of Phase One

Figure 6 depicts how light shelf height affects the daylighting performance of light shelves regarding the uniformity and distribution of WPI levels at distances of 1 m and 7 m from the window of the office where testing was performed. Remarkably, WPI levels in front of the window exceed 750 lux for all height levels used in this study and all design days. Additionally, the WPI levels of LS2H2 and LS3H3 are less compared to LS1H1, especially at midday. The illuminance level at the back daylit area was less than half of the illuminance level of the front area. However, the WPI levels were almost within the recommended range. Consequently, the uniformity index of LS1H1 is less than 0.5 at the winter solstice and during the mid-season due to the low angle of incidence of the Sun's path, as depicted in Figure 2. This leads to a higher contrast between daylight distribution near the window and the back surface. The LS2H2 and LS3H3 configurations achieved the uniformity index value at all times except during the winter solstice. It is worth mentioning that the width of the external light shelves influenced the integrated solar panel in terms of power generation during the last phase. However, its drawback is that there may be infringement concerns regarding the prospect right and damage due to wind pressure [6]. Therefore, the LS2H2 configuration was chosen for the width of the light shelf and 2/3rd of the window height, which is in line with previous studies [31].

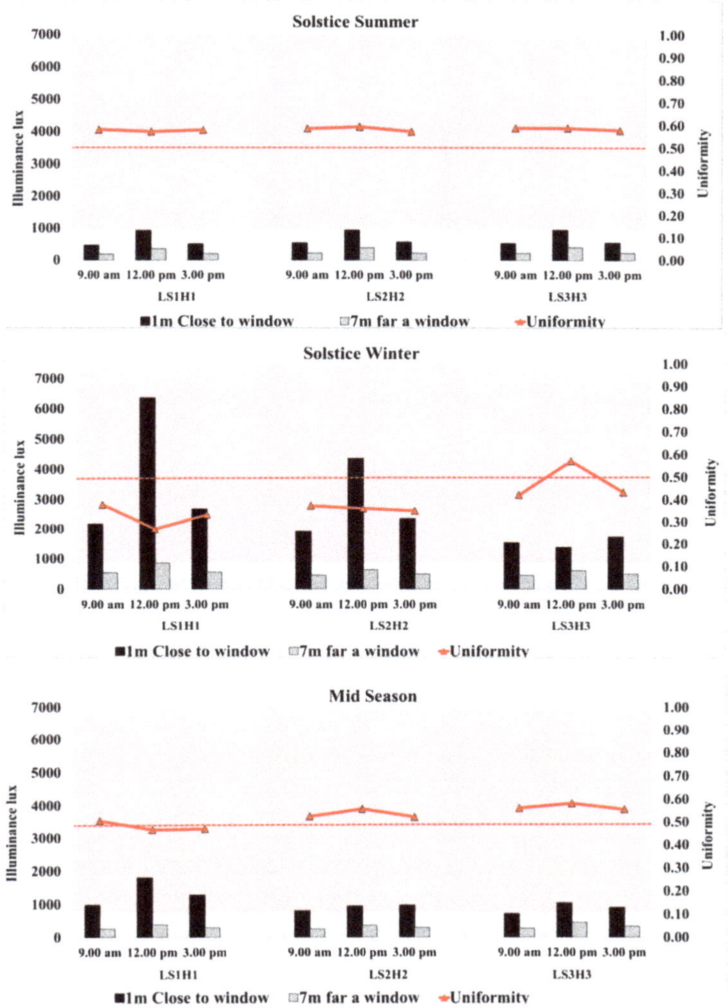

Figure 6. The effect of the position of internal and external light shelves at a distance of 1 m and 7 m from the window.

3.2. Analysis of Phase Two

Figure 7 depicts the introduction of the reflector material added at the top of the window compared to LS2H2. There is a remarkable increase of about 10% in WPI level distribution in the back daylit area. However, the WPI value pertaining to the front and back daylit area during the winter solstice is much higher than the recommended value, especially at midday. The uniformity index also witnessed improvement for all design days compared to the LS2H2 configuration (without a reflector), except for during the winter solstice. Thus, the use of a reflector at the top of the window, along with external and internal light shelves, is a useful daylight configuration for improving the WPI in the back space of the model. This result aligned with Zazzini et al.'s research [6]. Furthermore, uniformity is also enhanced in the said configuration; however, remarkable reflection is a concern when the Sun's path is low during the winter season.

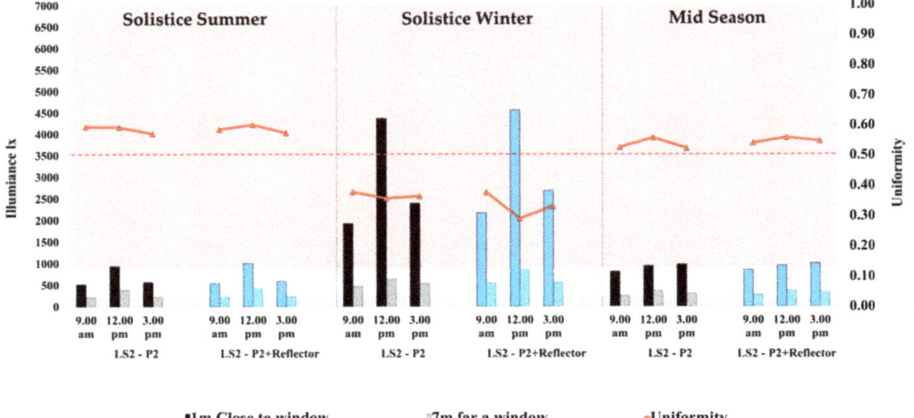

Figure 7. The performance of a reflector on the top of window combined with external and internal light shelves.

3.3. Analysis of Phase Three

Figure 8 depicts the distribution of indoor illuminance using three different internal curved LS angles (10%, 15%, and 25%) compared to horizontal LS on all design days. The results revealed that the use of an internal curved LS with 10° tilt angle recorded the highest value of WPI and uniformity index at a distance of 7 m from the window. The back daylit area saw an increase of up to 5–11% on these parameters during the summer and winter solstices; however, the increase was less than 4% at the spring equinox. The uniformity index also saw an improvement for all the seasons. The WPI value at about a 4 m distance from the window (middle area) and the front area is within the recommended range, except at midday, where it is consistently higher than 750 lux. The LSC3 configuration with 25° tilt angle presents the least optimal case since the WPI increased for the front area but decreased for the back area, which results in less uniformity in daylight distribution. Thus, the optimal tilt angle of LS is 10° (LSC1) for all seasons. In contrast, Heangwoo et al. [35] found that the optimal specifications for external curved light shelves are different depending on the season, but are effective at improving the indoor uniformity ratio compared to a flat light shelf during summer.

Overall, the observations from the optimized light shelf configurations after the three phases (height, reflector, curved internal light shelf) were compared with the reference model (without the light shelf). The proposed model illustrated a significant improvement considering the uniformity index on all design days, especially during midday. Furthermore, there was an increase of about 13–20% in the absolute WPI value concerning the back daylit area during the summer season for the entire day. At the same time, the absolute WPI value reduced for the middle (63°) and low (39°) sun path altitudes, as specified in Appendix Appendix B.

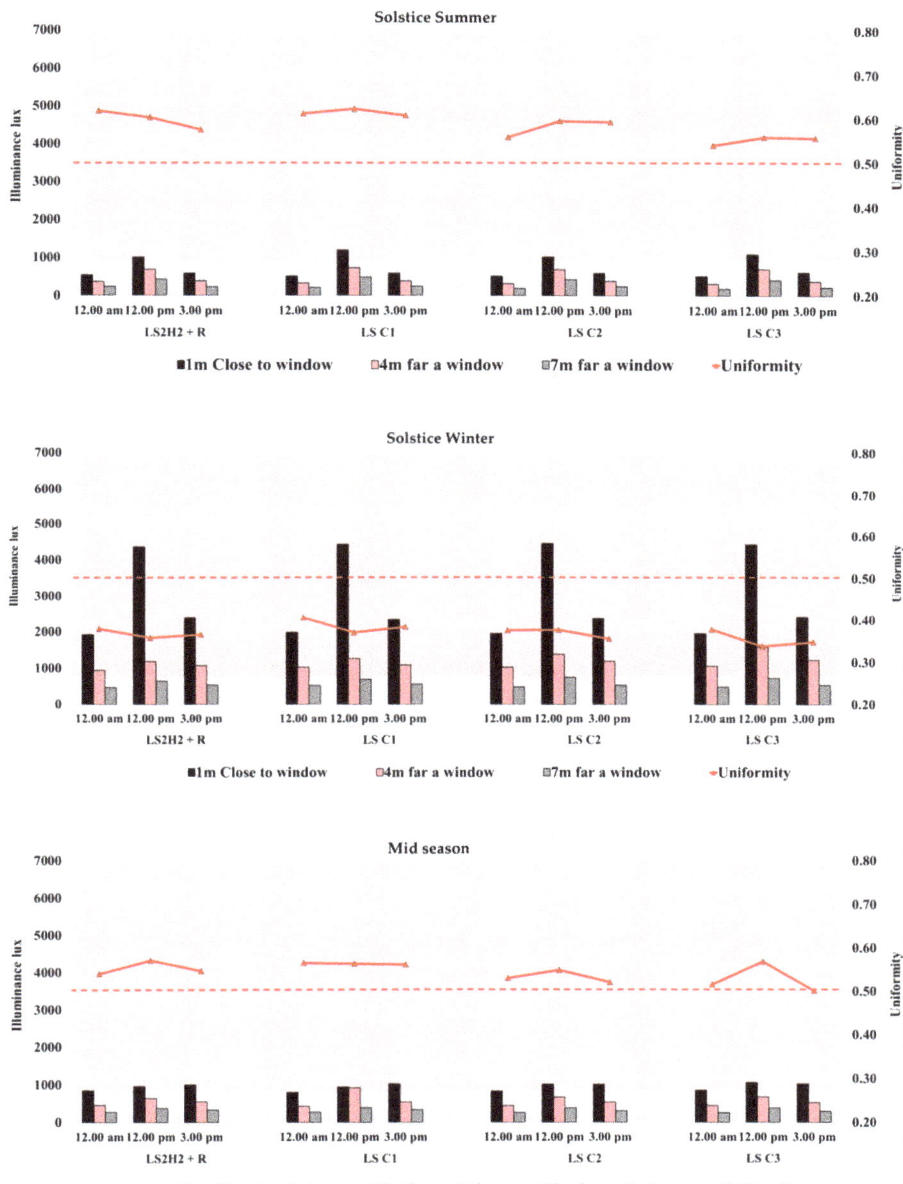

Figure 8. The performance of various curved light shelves in distance of 1 m, 4 m and 7 m from the window and Uniformity.

3.4. Performance Evaluation of Light Shelf Photovoltaic Configurations (LSPV) (Phase Four)

Tables 7–9 present the WPI distribution and uniformity index for the reference model and the many LSPV configurations corresponding to CIE clear sky within the design days. During the summer season, the WPI values and their averages concerning the reference model are higher than all LSPV configurations, especially during midday, as measured in the front area. It is also observed that in all cases, the uniformity index is higher than 0.5. Notably, the usage of LSPV leads to a noteworthy improvement in the WPI and uniformity index as compared to the reference model. The improvement in these parameters is better during midday, especially for configurations LSPV1 and LSPV2. In the case of the winter season, all WPI values are higher than 500 lux for all the configurations, except for the middle and back daylit area specific to the LSPV2 configuration. At the same time, the uniformity index improved for all the cases, but it was still less than 0.5. As for the mid-season results, the LSPV with a 30° tilted angle witnessed an increase in WPI at midday, while the LSPV with a horizontal angle witnessed an increase during the evening as a result of the low sun position. The only configuration that had uniformity was LSPV2, unlike the reference model.

Table 7. The point in time illuminance of PV integration with light shelves and uniformity at solstice summer.

Configurations	Solstice Summer	L1	L2	L3	L4	L5	L6	L7	Average	Uniformity
Reference Model	9.00 a.m.	441	432	363	292	236	196	177	305	0.58
	12.00 p.m.	1016	972	795	628	501	415	376	672	0.56
	3.00 p.m.	480	467	392	316	257	216	194	332	0.59
LSPV1	9.00 a.m.	352	333	278	229	190	162	148	242	0.61
	12.00 p.m.	658	644	543	449	376	323	300	470	0.64
	3.00 p.m.	427	408	351	299	259	230	215	313	0.69
LSPV2	9.00 a.m.	350	333	287	245	210	183	169	254	0.67
	12.00 p.m.	577	558	478	401	336	289	267	415	0.64
	3.00 p.m.	376	354	304	259	219	191	175	268	0.65
LSPV3	9.00 a.m.	380	362	303	248	205	175	159	262	0.61
	12.00 p.m.	764	755	645	530	441	372	341	550	0.62
	3.00 p.m.	426	404	399	278	230	197	180	302	0.60
LSPV4	9.00 a.m.	308	339	284	277	190	160	145	243	0.60
	12.00 p.m.	700	686	587	485	401	340	308	501	0.61
	3.00 p.m.	420	402	342	284	236	202	184	296	0.62
LSPV5	9.00 a.m.	472	454	385	320	265	226	205	332	0.62
	12.00 p.m.	849	838	711	583	480	404	368	605	0.61
	3.00 p.m.	562	537	460	385	323	279	254	400	0.64
LSPV6	9.00 a.m.	431	410	342	277	225	189	170	292	0.58
	12.00 p.m.	898	889	748	604	489	406	365	628	0.58
	3.00 p.m.	532	512	439	369	314	272	248	384	0.65
LSPV7	9.00 a.m.	502	481	405	337	268	225	203	346	0.59
	12.00 p.m.	932	918	771	627	513	428	390	654	0.60
	3.00 p.m.	565	546	466	385	321	276	260	403	0.65
LSPV8	9.00 a.m.	517	496	419	442	280	235	211	371	0.57
	12.00 p.m.	898	889	748	604	448	406	365	623	0.59
	3.00 p.m.	507	488	406	323	260	216	191	342	0.56

Table 8. The point in time illuminance of PV integration with light shelves and uniformity at solstice winter.

Configurations	Solstice Winter	L1	L2	L3	L4	L5	L6	L7	Average	Uniformity
Reference Model	9.00 a.m.	2434	3111	2501	1478	979	741	635	1697	0.37
	12.00 p.m.	7242	11280	8307	2669	1880	1420	1185	4855	0.24
	3.00 p.m.	2932	3881	3089	1816	1112	827	702	2051	0.34
LSPV1	9.00 a.m.	1804	2393	1571	914	638	515	450	1184	0.38
	12.00 p.m.	4166	2970	1520	1093	850	691	606	1699	0.36
	3.00 p.m.	2196	2197	1743	1002	670	543	474	1261	0.38
LSPV2	9.00 a.m.	1420	1927	1215	722	524	421	370	943	0.39
	12.00 p.m.	1703	1239	939	740	557	475	322	854	0.38
	3.00 p.m.	1635	2080	1250	763	546	435	380	1013	0.38
LSPV3	9.00 a.m.	1860	2235	1630	970	683	546	472	1199	0.39
	12.00 p.m.	4225	2990	1560	1117	880	712	621	1729	0.36
	3.00 p.m.	2228	2961	1498	1031	711	566	490	1355	0.36
LSPV4	9.00 a.m.	1550	2113	1336	784	551	445	391	1024	0.38
	12.00 p.m.	2497	1705	1139	882	688	562	499	1139	0.44
	3.00 p.m.	1854	2434	1458	870	620	485	420	1163	0.36
LSPV5	9.00 a.m.	1860	2446	1571	973	654	519	452	1211	0.37
	12.00 p.m.	4274	3080	1636	1190	922	750	654	1787	0.37
	3.00 p.m.	2308	3020	1817	1080	751	590	515	1440	0.36
LSPV6	9.00 a.m.	1728	2273	1518	881	630	500	432	1137	0.38
	12.00 p.m.	3234	2170	1340	1006	793	641	563	1392	0.40
	3.00 p.m.	2096	2695	1640	977	522	542	472	1278	0.37
LSPV7	9.00 a.m.	1943	2515	1686	994	707	560	484	1270	0.38
	12.00 p.m.	4332	3110	1662	1170	908	737	644	1795	0.36
	3.00 p.m.	2372	3085	1860	1127	800	633	551	1490	0.37
LSPV8	9.00 a.m.	1874	2464	1622	956	675	538	462	1227	0.38
	12.00 p.m.	3738	2585	1832	1067	825	670	586	1615	0.36
	3.00 p.m.	1773	2910	1096	1052	713	567	492	1229	0.40

Table 9. The point in time illuminance of PV integration with light shelves and uniformity at mid-season.

Configurations	Equinox Spring	L1	L2	L3	L4	L5	L6	L7	Average	Uniformity
Reference Model	9.00 a.m.	1208	856	645	502	394	324	284	602	0.47
	12.00 p.m.	6507	2525	1366	993	769	626	540	1904	0.28
	3.00 p.m.	2154	1173	814	615	481	394	348	854	0.41
LSPV1	9.00 a.m.	677	543	452	366	299	251	226	402	0.56
	12.00 p.m.	684	678	574	467	382	319	287	484	0.59
	3.00 p.m.	2196	2502	1743	1002	505	543	474	1281	0.37
LSPV2	9.00 a.m.	546	435	360	293	240	199	177	321	0.55
	12.00 p.m.	1703	1240	939	577	483	475	422	834	0.51
	3.00 p.m.	687	550	461	378	313	257	240	412	0.58
LSPV3	9.00 a.m.	704	580	478	386	312	259	231	421	0.55
	12.00 p.m.	799	789	676	551	447	374	336	567	0.59
	3.00 p.m.	2228	2960	1747	1031	711	566	490	1390	0.35
LSPV4	9.00 a.m.	652	534	441	361	295	252	228	395	0.58
	12.00 p.m.	2494	1700	1140	882	688	562	498	1138	0.44
	3.00 p.m.	825	653	554	454	373	315	282	494	0.57
LSPV5	9.00 a.m.	816	674	557	448	359	298	265	488	0.54
	12.00 p.m.	861	853	732	594	479	398	352	610	0.58
	3.00 p.m.	2308	3020	1817	1080	751	590	515	1440	0.36
LSPV6	9.00 a.m.	750	616	507	407	325	268	238	444	0.54
	12.00 p.m.	3234	2170	1037	1006	793	641	563	1349	0.42
	3.00 p.m.	897	741	600	482	388	322	288	531	0.54
LSPV7	9.00 a.m.	821	685	560	443	351	290	259	487	0.53
	12.00 p.m.	945	945	808	648	518	427	377	667	0.57
	3.00 p.m.	2322	3085	1860	1127	800	633	551	1483	0.37
LSPV8	9.00 a.m.	772	640	523	418	332	273	242	457	0.53
	12.00 p.m.	3738	2335	1460	1062	825	670	586	1525	0.38
	3.00 p.m.	933	757	619	493	382	318	280	540	0.52

Figure 9 presents the mean values of the DA and UDI metrics for the deep office considering all SPV configurations. The analysis revealed that the reference model achieved 78.9% of the 300 lux mean value of DA, which is the highest percentage compared to the LSPV configurations, where LSPV with 100% and 75% coverage is less than 50%. The annual distribution of DA for every grid point is depicted in Appendix Appendix C. In contrast, the UDI of LSPV (100–2000 lux) is higher at 90% compared to 83% for the reference model due to the high rate of visual discomfort compared to the LSPV configurations. The LSPV2 configuration almost eliminated visual discomfort (UDI > 2000).

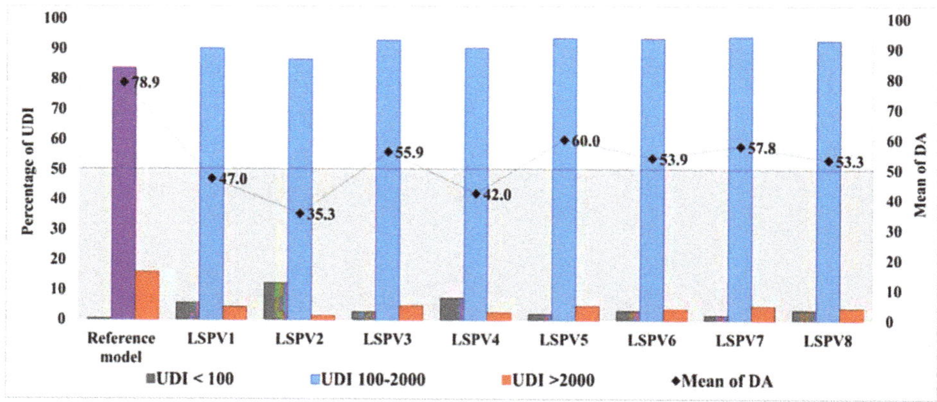

Figure 9. Annual climate-based analysis (UDI and mean of DA) of different configurations of light shelves integrated with photovoltaics.

Table 10 shows the magnitude of discomfort caused by glare for the reference model; optimal cases were determined using daylight glare probability (DGP) and a 3D contour map. The results revealed that a perceptible glare occurred only during the winter season when measured close to the window as per the reference model. This happened when the Sun was at its lowest and in direct view, specifically at noon. All LSPV configurations reduce the amount of DGP to an imperceptible glare condition. Eventually, configurations LSPV1 and LPSV2 with 100% coverage were considerably improved regarding visual comfort by reducing the mean values of DGP by at least nine degrees compared to the reference model.

Table 10. Illuminance contour maps of the reference model and the optimum LSPV configurations under clear sky at solstice winter at (9.00 a.m., 12.00 p.m., 3.00 p.m.).

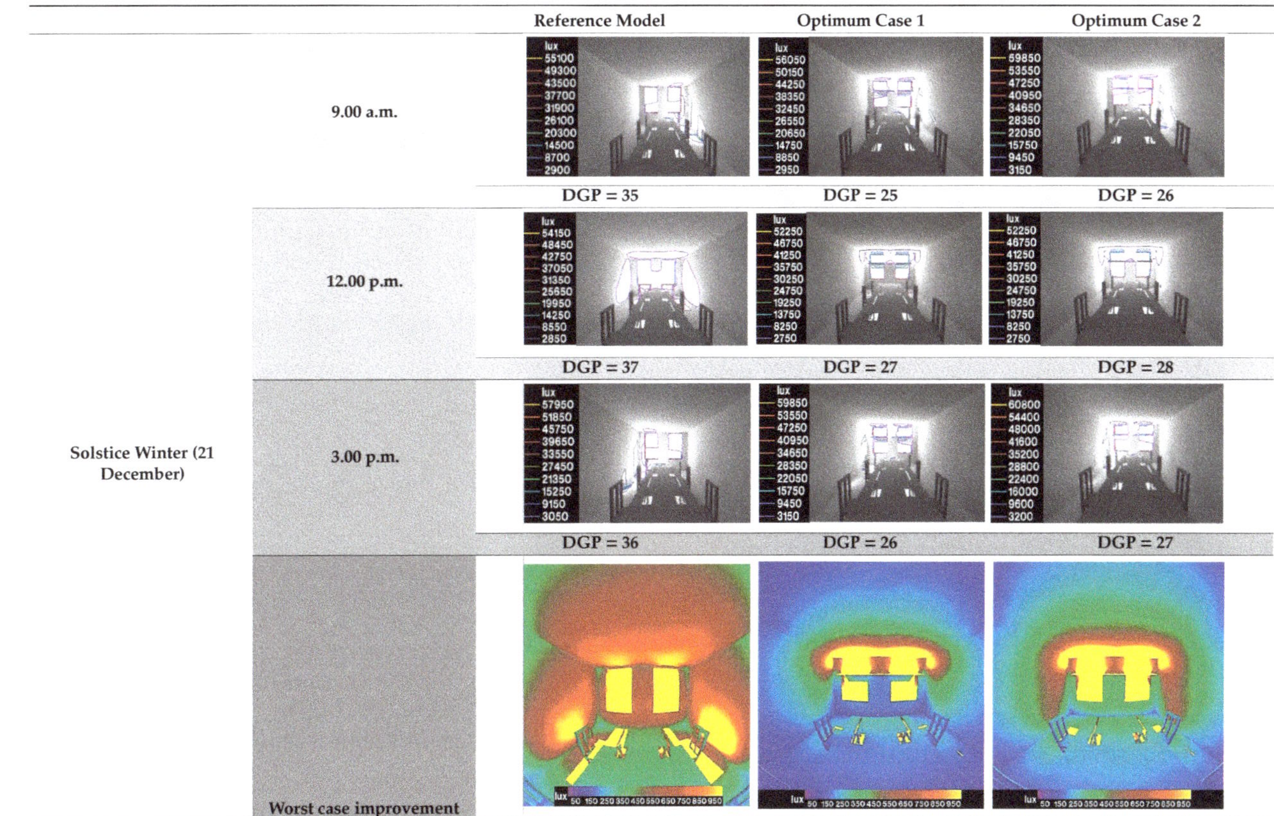

		Reference Model	Optimum Case 1	Optimum Case 2
Solstice Winter (21 December)	9.00 a.m.	DGP = 35	DGP = 25	DGP = 26
	12.00 p.m.	DGP = 37	DGP = 27	DGP = 28
	3.00 p.m.	DGP = 36	DGP = 26	DGP = 27
	Worst case improvement			

3.5. Analysis of Artificial Lighting Energy Consumption

Table 11 specifies the amount of energy production for the integrated LSPV and the total lighting energy consumption in each lighting control zone throughout the year pertaining to the climatic condition of Ha'il. For the analysis, it is assumed that lighting was switched on automatically after 8:00 a.m. and kept on until 5:00 p.m., with the dimming system set to 300 lux. The results show that the reference model (without LS) exhibited the lowest energy use compared to all LSPV configurations by a difference of only 20.6 kWh per year, while the lowest lighting energy consumption of 49.7 kWh was achieved for LSPV5, which is more than 50%. Peak lighting energy demand for the deep office reached up to 102 kWh for LSPV2.

Table 11. The yearly lighting energy consumption of various light shelf configurations compared to the base model.

Configuration of LS with PV	Lighting Energy Consumption (kWh)	PV Energy Production (kWh)	Net Energy Saving (kWh)	Percentage of Energy Saved with PV Light Shelves
Reference model	20.6	0	−20.6	0%
LSPV1	71.8	626	554	89%
LSPV2	102.9	686	583	85%
LSPV3	50.3	469	419	89%
LSPV4	93.8	514	420	82%
LSPV5	49.7	313	263	84%
LSPV6	53.6	343	289	84%
LSPV7	53.1	156	103	66%
LSPV8	63.1	171	108	63%

On the other hand, the energy output due to PV use was much higher than the lighting energy consumption. PV energy output ranged between 107 kWh for LSPV7 to 686 kWh for LSPV2. Consequently, the annual lighting-specific energy savings were calculated for two tilt angles (horizontal and 30°) and four PV coverages. The net savings were computed relative to the energy output of PV and lighting energy demands. In all the cases, the results indicated a reduction of at least 63% and, in some cases, up to 89%. Considering these observations, a light shelf using a solar module is advantageous concerning energy savings for lighting. Furthermore, the energy production can also compensate for a considerable fraction of other domestic energy needs, such as that required for cooling.

4. Conclusions

This study focused on the application of using integrated photovoltaic solar panels in light shelves to decrease the lighting energy requirement for office buildings and proposed a prototype of a modular unit composed of a light shelf combined with photovoltaic technology (LSPV) for deep office buildings in hot desert-like climatic conditions. In order to optimize the daylighting performance, three phases were carried out before the PV was attached to the LS to determine the appropriate height, reflector characteristics, and curved internal LS. The key findings of this study are specified below:

- The optimal height for a flat LS for enhancing WPI and the uniformity index determined in this study is 1.3 m above the floor with widths of 1.1 m and 0.7 m for the external and internal LS respectively, Configuration LS3H3 is challenging to use in high rise buildings.
- The use of a reflector with constant width of 30 cm at the top of the window, combined with external and internal LS configurations, is considered a good daylight strategy to improve the WPI in the back area of the model by 10%.
- The optimal specification for a flat external LS, combined with a curved internal LS, for improving daylighting distribution was found to be 10°, which increased the back daylit area by up to 5–11% at the summer and winter solstices, and less than 4% at the spring equinox.
- All LS configurations are observed to provide less daylight in the back daylit area compared to the reference model; however, there is uniformity of illumination.

- The integration of LSPV with a conversion efficiency of only 10% can completely compensate for the lighting-specific energy consumption in all LSPV configurations. At the same time, these configurations eliminate the discomfort caused due to glare, especially during the winter season. Also, LSPV1 with a flat and 30° tilt angle with 100% coverage shows a higher uniformity of illumination compared to a light shelf without a solar module.
- The optimal modular units of the LSPV that can achieve significantly greater savings and uniformity index within the office perimeter, close to the windows and the middle, along with the back area, are LSPV1 and LSPV2, as specified in Appendix Appendix D.

The focus of this study is limited to the energy required for lighting and daylight distribution for visual comfort. However, further studies are needed to evaluate the effect of the wall-to-window ratio (WWR) and other types of photovoltaic materials, specifically with respect to conversion efficiency, the effects of transparency on energy saving, and glare prevention. The effects of combined internal and external light shelves on window view also need further investigation. Moreover, the potential integration of LSPV had a significant impact on providing uniform daylight and preventing CO_2 emissions. Finally, such structures can be conveniently installed in buildings during renovation.

Author Contributions: Conceptualization, A.M. and A.G.; methodology, A.M.; software, A.M..; validation, A.M.; formal analysis, A.M.; investigation, A.M.; resources, A.M.; data curation, A.M. and A.G.; writing—original draft preparation, A.M.; writing—review and editing, A.G.; visualization, A.G.; supervision, A.G.; project administration, A.G.; funding acquisition, A.G. All authors have read and agreed to the published version of the manuscript.

Funding: This research received no external funding.

Acknowledgments: I would like to thanks Mesloub said for his support to complete this project. Also the authors would thank the editor, the managing editor, the academic editor, and three anonymous reviewers for their helpful comments which improved earlier versions of manuscript.

Conflicts of Interest: The authors declare no conflict of interest.

Nomenclature

LSPV	Light shelf photovoltaics
ILS	Internal light shelf
ELS	External light shelf
STPV	Semi-transparent photovoltaics
ISO	International Organization for Standardization
CEI	Commission International d'Eclairage
WPI	Work plane illuminance
TMY	Meteonorm metrological database
LS1H1	Light shelf height
LSC	Light shelf curved
LS2H2 + R	Light shelf height + reflector
VLT	Visible light transmittance
WWR	Window-to-wall ratio
UI	Uniformity index
UDI	Useful daylight illuminance
DA	Daylight autonomy
DGP	Daylight glare index
$h_{clerestory}$	Height of clerestory
$d_{int\ light\ shelf}$	Depth of internal light shelf
$d_{ext\ light\ shelf}$	Depth of external light shelf

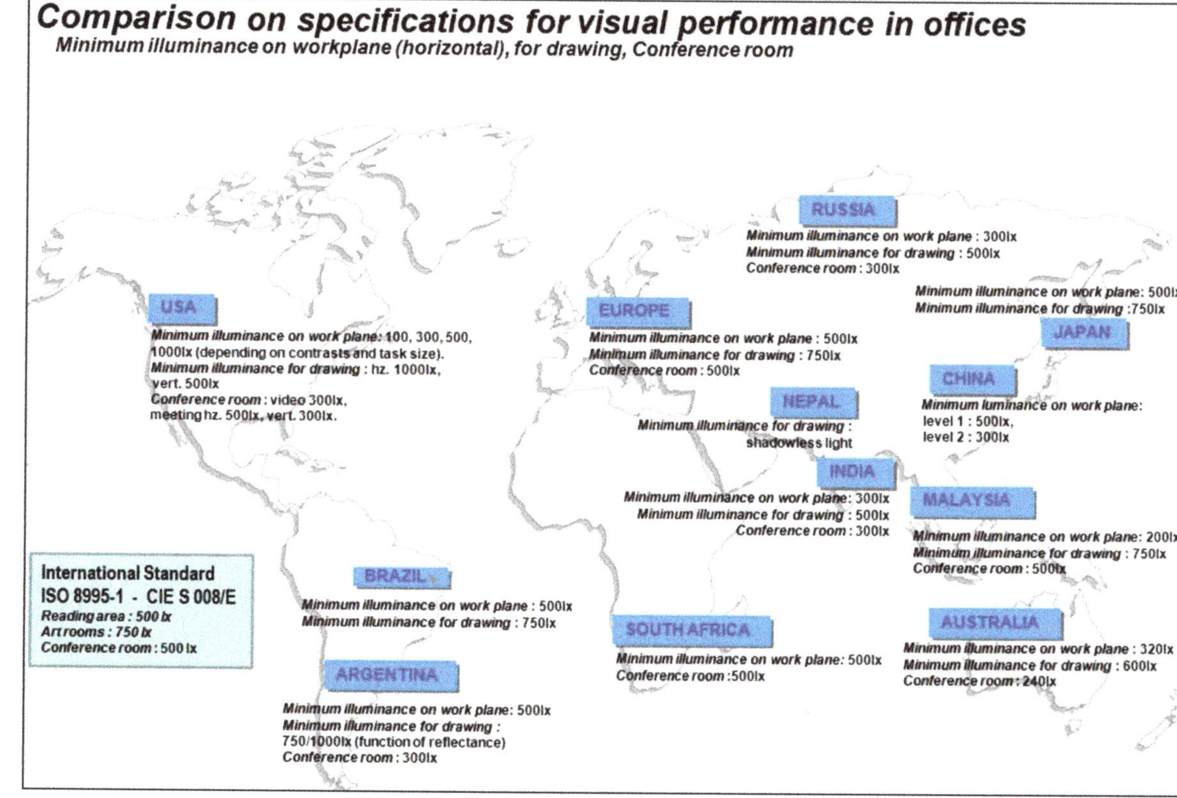

(A)

Figure A1. *Cont.*

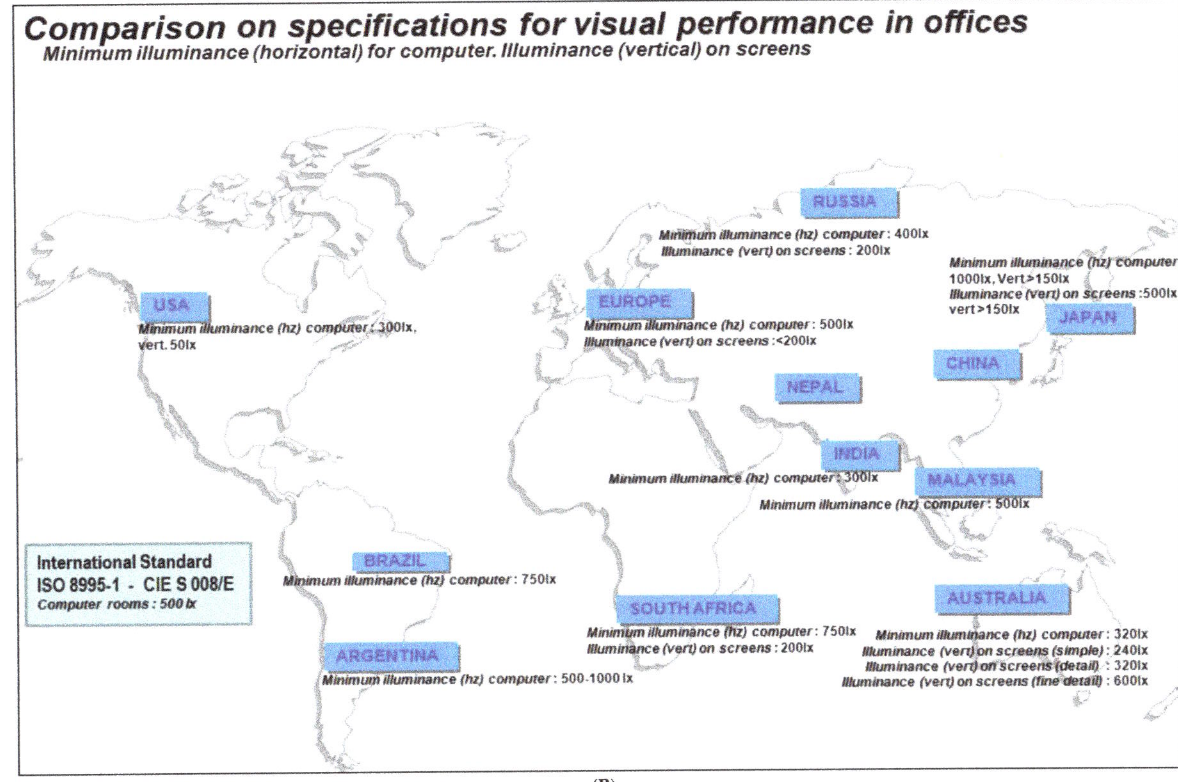

Figure A1. The minimum illuminance level based on International Standard ISO 8995-1:2002 (CIE 2001/ISO 2002) for visual performance in office (**A**) WPI horizontal for a drawing task (**B**) for a computer task.

Appendix B.

Table A1. The distribution of absolute work plan illuminance (WPI) and uniformity of the reference model compared to the optimized internal and external light shelves without integrating a solar module in (**a**) solstice summer, (**b**) equinox spring, and (**c**) solstice winter.

(a)

Configurations	Solstice Summer	L1	L2	L3	L4	L5	L6	L7	Average	Uniformity
Reference Model	9.00 a.m.	441	432	363	292	236	196	177	305	0.58
	12.00 p.m.	1016	972	795	628	501	415	376	672	0.56
	3.00 p.m.	480	467	392	316	257	216	194	332	0.59
LS C1	9.00 a.m.	511	490	409	332	270	226	214	350	0.61
	12.00 p.m.	1196	1039	893	736	612	524	488	784	0.62
	3.00 p.m.	593	572	483	393	320	269	250	411	0.61

(b)

Configurations	Equinox Spring	L1	L2	L3	L4	L5	L6	L7	Average	Uniformity
Reference Model	9.00 a.m.	1208	856	645	502	394	324	284	602	0.47
	12.00 p.m.	6507	2525	1366	993	769	626	540	1904	0.28
	3.00 p.m.	2154	1173	814	615	481	394	348	854	0.41
LS C1	9.00 a.m.	797	715	546	436	347	286	275	486	0.56
	12.00 p.m.	943	935	791	923	494	402	393	697	0.56
	3.00 p.m.	1023	841	690	550	437	362	340	606	0.56

(c)

Configurations	Solstice Winter	L1	L2	L3	L4	L5	L6	L7	Average	Uniformity
Reference Model	9.00 a.m.	2434	3111	2501	1478	979	741	635	1697	0.37
	12.00 p.m.	7242	11,280	8307	2669	1880	1420	1185	4855	0.24
	3.00 p.m.	2932	3881	3089	1816	1112	827	702	2051	0.34
LS C1	9.00 a.m.	1984	2569	1747	1023	731	586	531	1310	0.41
	12.00 p.m.	4432	3210	1760	1267	985	801	695	1878	0.37
	3.00 p.m.	2338	3066	1866	1091	769	610	564	1472	0.38

Appendix C.

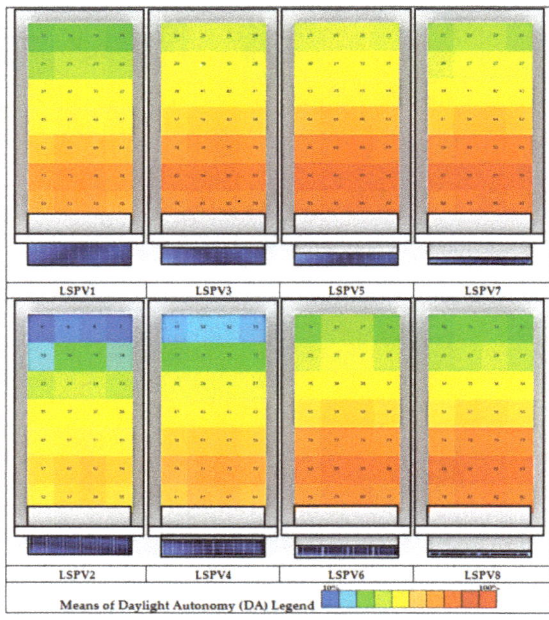

Figure A2. Daylight autonomy (DA) distribution of different LSPV configurations in each row.

Appendix D.

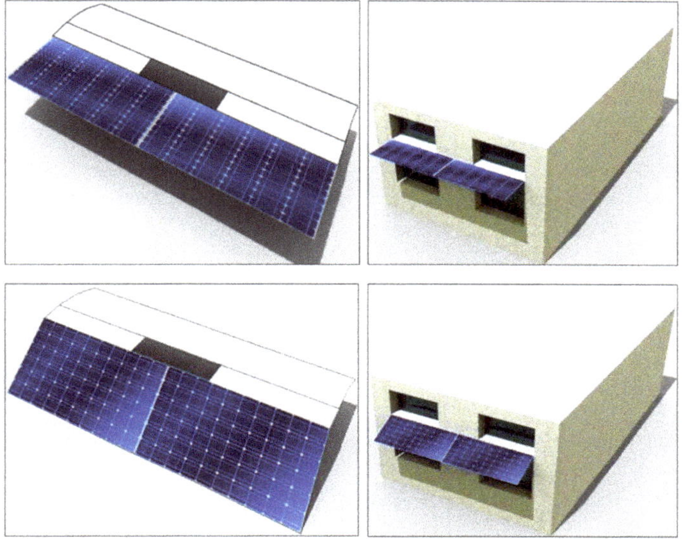

Figure A3. The optimum LSPV configuration design.

References

1. Saudi Vision 2030. Available online: https://vision2030.gov.sa/en/node/6 (accessed on 5 November 2020).
2. Qahtan, A.M.; Ebrahim, D.A.; Ahmed, H.M. Energy-saving potential of daylighting in the Atria of Colleges in Najran University, Saudi Arabia. *Int. J. Built Environ. Sustain.* **2020**, *7*, 47–55. [CrossRef]
3. Zell, E.; Gasim, S.; Wilcox, S.; Katamoura, S.; Stoffel, T.; Shibli, H.; Engel-Cox, J.; Al Subie, M. Assessment of solar radiation resources in Saudi Arabia. *Sol. Energy* **2015**, *119*, 422–438. [CrossRef]
4. Efficiency, S.E. *Saudi Energy Efficiency Program*; Governmental and Commercial Sector: Riyadh, Saudi Arabia, 2018.
5. Lim, Y.W.; Heng, C. Dynamic internal light shelf for tropical daylighting in high-rise office buildings. *Build. Environ.* **2016**, *106*, 155–166. [CrossRef]
6. Zazzini, P.; Romano, A.; di Lorenzo, A.; Portaluri, V.; di Crescenzo, A. Experimental analysis of the performance of light shelves in different geometrical configurations through the scale model approach. *J. Daylighting* **2020**, *7*, 37–56. [CrossRef]
7. Kontadakis, A.; Tsangrassoulis, A.; Doulos, L.; Zerefos, S. A review of light shelf designs for daylit environments. *Sustainability* **2018**, *10*, 71. [CrossRef]
8. Mesloub, A.; Albaqawy, G.A.; Kandar, M.Z. The OPTIMUM performance of Building Integrated Photovoltaic (BIPV) windows under a semi-arid climate in algerian office buildings. *Sustainability* **2020**, *12*, 1654. [CrossRef]
9. Gutiérrez, R.U.; Du, J.; Ferreira, N.; Ferrero, A.; Sharples, S. Daylight control and performance in office buildings using a novel ceramic louvre system. *Build. Environ.* **2019**, *151*, 54–74. [CrossRef]
10. Heng, C.; Lim, Y.W.; Ossen, D.R. Horizontal light pipe transporter for deep plan high-rise office daylighting in tropical climate. *Build. Environ.* **2020**, *171*, 106645. [CrossRef]
11. Freewan, A.A. Maximizing the lightshelf performance by interaction between lightshelf geometries and a curved ceiling. *Energy Convers. Manag.* **2010**, *51*, 1600–1604. [CrossRef]
12. Chieli, G.; Nelli, L.C. Photovoltaic and thermal solar concentrator integrated into a dynamic shading device. In *Sustainable Building for a Cleaner Environment*; Springer: Berlin, Germany, 2019; pp. 335–345.
13. Lee, H. Performance evaluation of a light shelf with a solar module based on the solar module attachment area. *Build. Environ.* **2019**, *159*, 106161. [CrossRef]
14. Raphael, B. Active control of daylighting features in buildings. *Comput. Aided Civ. Infrastruct. Eng.* **2011**, *26*, 393–405. [CrossRef]

15. Meresi, A. Evaluating daylight performance of light shelves combined with external blinds in south-facing classrooms in Athens, Greece. *Energy Build.* **2016**, *116*, 190–205. [CrossRef]

16. Lee, H.; Jang, H.I.; Seo, J. A preliminary study on the performance of an awning system with a built-in light shelf. *Build. Environ.* **2018**, *131*, 255–263. [CrossRef]

17. Hwang, T.; Kim, J.T.; Chung, Y. Power performance of photovoltaic-integrated lightshelf systems. *Indoor Built Environ.* **2014**, *23*, 180–188. [CrossRef]

18. Berardi, U.; Anaraki, H.K. Analysis of the impacts of light shelves on the useful daylight illuminance in office buildings in Toronto. *Energy Procedia* **2015**, *78*, 1793–1798. [CrossRef]

19. Moazzeni, M.H.; Ghiabaklou, Z. Investigating the influence of light shelf geometry parameters on daylight performance and visual comfort, a case study of educational space in Tehran, Iran. *Buildings* **2016**, *6*, 26. [CrossRef]

20. Kim, K.; Lee, H.; Jang, H.; Park, C.; Choi, C. Energy-saving performance of light shelves under the application of user-awareness technology and light-dimming control. *Sustain. Cities Soc.* **2019**, *44*, 582–596. [CrossRef]

21. Lee, H.; Park, S.; Seo, J. Development and performance evaluation of light shelves using width-adjustable reflectors. *Adv. Civ. Eng.* **2018**, *2018*, 2028065. [CrossRef]

22. Bahdad, A.A.S.; Fadzil, S.F.S.; Taib, N. Optimization of daylight performance based on controllable light-shelf parameters using genetic algorithms in the tropical climate of malaysia. *J. Daylighting* **2020**, *7*, 122–136. [CrossRef]

23. Lee, H.; Seo, J. Performance evaluation of external light shelves by applying a prism sheet. *Energies* **2020**, *13*, 4618. [CrossRef]

24. International standard organization(ISO). *Lighting of Indoor Work Places*; ISO: Geneva, Switzerland, 2002.

25. DiLaura, D.L. Illuminating Engineering Societythe Lighting HandbookTenth Edition Reference and Application. 2011. Available online: http://ndl.ethernet.edu.et/handle/123456789/23109 (accessed on 9 November 2020).

26. JISZ9110. *Recommended Levels of Illumination*; Japanese Industrial Standards Committee: Tokyo, Japan, 2010.

27. De Normalisation, C.E. *EN 12464-1: Light and Lighting-Lighting of Work Places, Part 1: Indoor Work Places*; Comité Européen de Normalisation: Bern, Switzerland, 2002.

28. Bellia, L.; Pedace, A.; Fragliasso, F. Dynamic daylight simulations: Impact of weather file's choice. *Sol. Energy* **2015**, *117*, 224–235. [CrossRef]

29. Remund, J.; Müller, S.; Kunz, S.; Huguenin-Landl, B.; Studer, C.; Klauser, D.; Schilter, C.; Lehnherr, R. Meteonorm global meteorological database. In *Hanbook I/II*; Meteotest: Bern, Switzerland, 2013.

30. Berardi, U.; Anaraki, H.K. The benefits of light shelves over the daylight illuminance in office buildings in Toronto. *Indoor Built Environ.* **2018**, *27*, 244–262. [CrossRef]

31. Joarder, M.; Rahman, A.; Ahmed, Z.N.; Price, A.; Mourshed, M. *A Simulation Assessment of the Height of Light Shelves to Enhance Daylighting Quality in Tropical Office Buildings under Overcast Sky Conditions in Dhaka, Bangladesh*; International Building Performance Simulation Association: Glasgow, UK, 2009.

32. Hu, J.; Du, J.; Place, W. The assessment of advanced daylighting systems in multi-story office buildings using a dynamic method. In Proceedings of the World Renewable Energy Congress-Sweden, Linköping, Sweden, 8–13 May 2011; pp. 1867–1874.

33. Kensek, K.; Suk, J.Y. Daylight factor (overcast sky) versus daylight availability (clear sky) in computer-based daylighting simulations. *J. Creat. Sustain. Archit. Built Environ.* **2011**, *1*, 3–14.

34. Abdelhakim, M.; Lim, Y.W.; Kandar, M.Z. Optimum glazing configurations for visual performance in algerian classrooms under mediterranean climate. *J. Daylighting* **2019**, *6*, 11–22. [CrossRef]

35. Lee, H.; Seo, J.; Choi, C.H. Preliminary study on the performance evaluation of a light shelf based on reflector curvature. *Energies* **2019**, *12*, 4295. [CrossRef]

Publisher's Note: MDPI stays neutral with regard to jurisdictional claims in published maps and institutional affiliations.

Article

The Spatial Distribution and Influencing Factors of Employment Multipliers in China's Expanding Cities

Daquan Huang [1], Han He [1] and Tao Liu [2,3,*]

[1] School of Geography, Faculty of Geographical Science, Beijing Normal University, No. 19, XinJieKouWai St., HaiDian District, Beijing 100875, China; huangdaquan@bnu.edu.cn (D.H.); 201821051048@mail.bnu.edu.cn (H.H.)
[2] College of Urban and Environmental Sciences, Peking University, Yiheyuan Road 5, Beijing 100871, China
[3] Center for Urban Future Research, Peking University, Yiheyuan Road 5, Beijing 100871, China
* Correspondence: liutao@pku.edu.cn

Abstract: In the process of urbanization in developing countries, creating enough jobs to realize the transition from an agricultural population to a non-agricultural population is a major goal of development. The differences and localities of cities need to be considered in the policymaking process. This study estimated the local employment multipliers of expanding cities in China and calculated the employment multiplier of each city. First, there are obvious differences in the size of employment multipliers across cities; therefore, it is necessary to adopt different policies in employment promotion. Second, an inverted-U-shape relationship is detected between employment multiplier and city size, namely the larger the city, the greater the employment multiplier, but when the city size exceeds a certain value, the employment multiplier begins to decline. Third, different degrees of influence are generated by factors for cities at different levels of economic development. Based on the research results, we suggest that expansion of the trade sector be promoted in small- and medium-sized cities, to give full play to its employment multiplier effect; meanwhile, in large cities, the degree of specialization of the trade sector and diversification of the non-trade sector should be improved.

Keywords: expanding cities; employment multiplier; spatial distribution; influencing factors; China

Citation: Huang, D.; He, H.; Liu, T. The Spatial Distribution and Influencing Factors of Employment Multipliers in China's Expanding Cities. *Appl. Sci.* **2021**, *11*, 1016. https://doi.org/10.3390/app11031016

Received: 30 December 2020
Accepted: 21 January 2021
Published: 23 January 2021

1. Introduction

Problems associated with urban employment have received increasing attention; in particular, fluctuations in the international trade environment and regional competition among cities have led to an alarming degree of job flux between cities [1,2]. While some cities have excellent employment vitality, others have sustained job losses and rising unemployment. William Julius Wilson considered the issue of unemployment as the core and origin of sundry American urban pathologies [3]. This problem is not exclusive to the United States; developing countries face the same issues, thus aiming for much higher employment and the resolution of unemployment [1,4–6]. Scholars have analyzed the problem of employment in developing countries and the significance of employment promotion [7]. First, job creation is key to realizing the shift of labor from the low-income and low-outcome sectors to the high-income and high-outcome sectors in developing countries. It is also directly related to the reduction in the size of unemployment. Second, increasing opportunities for the poor to obtain jobs and better their lives would address income redistribution. Even though there is some dispute over the reasons for job creation—for example, division among economists on whether productivity and marginal product growth are consistent with job creation—and on issues of efficiency and equity [7], efforts to create jobs have always been a relevant topic among academics and governments. The importance of employment varies from country to country, at different levels of development. Even in developing countries, there are significant differences between different institutional systems.

China has made remarkable achievements in urbanization in the 21st century. In the future, China will continue to promote urbanization, which means that a large proportion of the agricultural labor force will be converted into a non-agricultural labor force [8]. Lewis's classic paper on the labor market [9] concluded that one of the core functions of Chinese cities is to affect the spatial transfer of a large proportion of the agricultural population from low-income and low-output sectors to high-income and high-yield sectors in the process of urbanization; this means that how to respond to the increase in urban employment demand is a practical problem that Chinese cities have needed to face and solve for a long time.

When it comes to how to create more jobs, economists have validly offered a series of classic suggestions, including using the basic multiplier [10,11]. Economists argue that the basic activities of a local economy are premised on the existence and development of non-basic activities and that a city's service sector will decline if jobs are lost in the trade sector. This is true, regardless of how well developed a city's service sector currently is [12], such as in the case of Detroit. Therefore, the local government is committed to attracting enterprises and productive activities through a series of incentive measures [13]. It is believed that such immigration will not only bring a direct increase in employment and income but also have a driving effect on other local industries through a multiplier effect [14]. Specifically, this facilitation effect, known as the "local multiplier effect", is mainly achieved through the demand of labor forces in the tradable sector for production and life services of the local non-tradable sector [15].

The concept of the multiplier originates from Keynesian economics [15]. In the 1960s and 1970s, some scholars tried to advance the idea of an employment multiplier [16] and used it to study the economic base multiplier [10]. However, the idea did not receive much attention at the time. Until the early 2000s, Moretti modeled and estimated the size of America's overall employment multiplier [11]. Subsequently, this model has been used to estimate the employment multiplier in several countries and regions, including Japan [17], Sweden [14,18], Italy, and the United States [11,19–21]. As a result, a wealth of conclusions was obtained in a short time. Scholarship on the employment multiplier has confirmed its practical significance and discovered significant differences between different regions. These differences are also reflected in the choices of spatial scale and research time [22,23].

In addition to the selection of the time and spatial scales, the multiplier effects of different positions in the trade industry also differ. It is generally believed that highly technical positions bring more multipliers [14,24]. As for the formation of differences in employment multipliers within the industry, neo-Keynesian economists believe that highly skilled individuals and households will have higher consumer spending because they have higher wage incomes [11,25], resulting in more demand for non-tradable products and services. From the perspective of upstream and downstream industrial structure and production, higher technical levels will increase the likelihood [26] that not only will each additional job in the upstream end of the production system result in immediate employment demand, but the trade department will also bring jobs to the middle and lower reaches of production activities. However, the employment multipliers for skilled jobs of the same levels in different cities also differ [24], which shows the importance of considering differences among cities when discussing the multiplier.

In addition to the estimation of the employment multiplier, scholars have explained the reasons for differences in multipliers from different perspectives. The present study attempts to explain the reason for these differences but offers mostly descriptive explanations of single factors, such as urban scale [27], development level and location [22], and mobility [17]. Some research model analyses have also examined the relationship between individual factors and employment multipliers. All the above provide many theoretical and empirical foundations for understanding the multiplier. Given the background of new economic geography, production trade and labor mobility between urban areas are important for understanding a city's employment. With this approach, the multiplier theory of overflow and model is used to explore the preliminary results [28]. Recent empirical studies incorporate the spatial spillover of the employment effect into the study of employment multipliers [19].

There are deficiencies in the literature. First, both the existence and differences of employment multipliers have been confirmed [11,22,23], but the research scale has been concentrated at the macro regional and national levels. In terms of policy suggestions for urban development, a differentiation policy is necessary. Different cities have shown huge differences in city size, population structure, function, and locational conditions; these difference are particularly evident in developing countries, including China, which is undergoing rapid urbanization. Hence, this necessitates different development policies [29]; for example, development policies that work well in big cities may not have the same effect in smaller ones [30]. The small-city employment multiplier has been studied with this in mind [27]. However, the differences between cities are not just about size. On this basis, first, we need to estimate the employment multiplier of each city and the reasons for the difference in order to formulate effective employment promotion policies. Second, a mechanism analysis of the differences among employment multipliers is lacking. Most research addresses applications and tests of the model proposed by Moretti [11]. Almost all scholars have affirmed the differences in employment multipliers between industries, along different spatial scales, and in different cities [19,20,22,23], but they have not conducted further studies on the theoretical framework of the formation of differences. Third, Pred and Krugman argue that cities have different employment multipliers at different stages of development [31], especially during expansion and recession. While all prefectural cities are included in the study of China [22], including shrinking cities, the actual multiplier of urban sprawl is understated. Although the regionalization approach gave greater depth to the multiplier, the error generated by including shrinking cities could not be completely eliminated.

This study identified Chinese prefecture-level cities in the expansion process. The employment multiplier for these cities was estimated by using the City Statistical Yearbook data and OLS (Ordinary Least Square) regression. In addition to calculating the urban employment multiplier, this paper establishes a basic understanding of the framework and analyzes the formation mechanism of such differences within this framework to understand the causes of the differences, based on which development suggestions are provided. Section 2 establishes an analytical framework that combines relevant research. The methods of the study and resulting data are discussed in Section 3. Section 4 introduces the urban employment multiplier calculation results and the space–time evolution law. The paper concludes with a discussion and suggestions.

2. Analytical Framework

The expression "employment multiplier" refers to the speed and effect of change in the non-tradable sector brought about by the change in the trade sector. Therefore, it includes both positive and negative aspects. However, most of the multipliers mentioned in the current literature show the positive promotion effect brought about by the growth of the trade or manufacturing sector [23,28,32]. Krugman argues that the employment multiplier goes beyond a single positive effect to take a more interesting and complex form. This effect is mainly reflected in the process of urban contraction and recession [33]. When jobs in the trade sector are lost, the change rate of total urban employment is different from that of non-tradable sector employment. Economists believe that the rate of change caused by employment increases and decreases in the non-tradable sector is different from that of the trade sector [31]. There are different degrees of effects in creating and reducing trade sector jobs of the same scale within the same city, which means that incorporating all cities into one model would skew the results of the employment multiplier in the presence of shrinking cities.

The article combines the multiplier theory with the theory of the urban development stage, and offers the phased view that increases in the trade sector during urban expansion will bring about a significant employment multiplier effect. This is consistent with the results of most studies [20,21,27]. When the expansion of a city has gradually slowed to a steady state, the growth rate of the trade sector also slowed. In this state of development, the non-tradable sector will no longer grow in multiplier form, and the two sectors may show a negative relationship in terms of growth rate. When a city enters a recession, the tradable and non-tradable sectors are simultaneously reduced. The decline of the tradable sector then leads to the contraction of the non-tradable sector at a different rate than when cities are expanding. As a result, this study found it necessary to distinguish between expanding and shrinking cities in order to calculate a more precise multiplier effect of employment.

Joan Robinson proposed a theoretical framework to inform our understanding of the urban employment multiplier effect in China. Robinson constructed the formation process and influencing factors of the multiplier from the perspective of economics [34]. She argued that the employment multiplier is affected in two ways. The first is the number of times income passes through different cycles, which is the geometric progression. The second is the degree of residents' consumption, which is reflected by the wage–profit ratio, relief–wage ratio, and savings–profit ratio. Robinson's contribution was to construct a complete dynamic analytical framework of employment multipliers from individuals to cities to countries, rather than simply finding the influencing factors. This is of great help for us to analyze the employment multiplier in China, but Robinson's framework is based on the capitalist market economy, and there are differences in how influencing factors express themselves in Chinese cities.

A basic theory of the formation of the multiplier is established based on the theory proposed by Robinson [34]. The realization of the employment multiplier effect from the trade sector to the local non-trade sector is based on the consumption demand of the production and life of the practitioners in the tradable sector [11,15]. Manufacturing sector workers earn wage income, which is used for production and consumption, creating effective demand as a result. New Keynesian economics takes the perspective that in a capitalist market, one part of a household's wages is spent while the rest is saved. The consumption aspect influences the urban labor market in two ways. The first part consists of the demands of the entire trade department for related products and services, including innovation, design, advertisement, and law. The second aspect consists of influencing the urban services industry through individuals, such as catering and entertainment. These two types of consumer demand lead to the first step in multiplier realization. The jobs generated by the first approach pay higher wages, which Alan Scott and others call "cultural cognition". The second method produces low-income jobs, which belong to the category of physical practice. The increase in these jobs will bring income to the corresponding job recipients; these recipients will also consume in the city, thus further generating the second-step employment multiplier, which is generally smaller than the first-step multiplier. The second-step multiplier is also affected by individual income: The higher the income, the larger the scale of consumption and the higher the employment multiplier. In the same way, a multi-step employment multiplier is generated, and the total employment multiplier is finally obtained through addition. This means that a higher-wage income sector will generally correlate with a higher employment multiplier [20]. The cyclical coherence of employment and public spending has also been demonstrated in several countries [35].

The relationship between cities is the key to the size of the urban employment multiplier [19]. The previous idea was based on the assumption that the region has no connection to the outside world. In other words, the employment effect produced by the trade sector is realized in the city. In reality, cities are related to other regions as components of urban systems. Some of the local consumption spending spills over to other cities through this relationship, so part of the employment multiplier from the increase in the trade sector spills over to the linked cities [36], as scholars believe that export-oriented strategies are less effective in promoting employment than import substitution [37]. However, the spillover effects between cities are not equal. According to the theory of centrality, it is not difficult to deduce that in urban networks, the employment overflow in big cities is much higher than that in small cities [38,39]. The close relationships between cities makes the transformation of spatial scale very important in the process of understanding and analyzing the urban employment multiplier. Keynes also discusses the differences in employment multipliers across different scales in his work [15].

Scholars have also proposed factors influencing the employment multiplier from the perspective of geography. Moretti identified three factors affecting the urban employment multiplier [11,14]. First, the employment multiplier is directly related to consumers' consumption preferences for non-traded goods. As mentioned above, the realization of the employment multiplier is based on residents' consumption activities. If residents save more of their income, the employment demand for non-tradable sectors in the city will decrease, thus keeping the employment multiplier at a low level. Second, the type of employment in the trade sector also has important implications because higher income leads to greater growth in demand for local services [24]. Third, the labor-and-housing-supply elasticity, which affects the space overflow of economic activities. If the supply is completely elastic, the multiplier is the largest.

Economists and geographers have achieved much in employment multiplier theory and empirical research, but not enough to explain the differences in employment multipliers, especially to explain the horizontal differences between cities within the same economic system. Based on scholarly research, and considering the particularity of China as a large developing country, this paper determines the following factors of concern. The first is the economy and industrial structure. Both the specialization degree of the trade sector and the diversification degree of the non-trade sector have important internal relationships with urban employment. The former determines the income levels of the trading sector and the city as a whole [12,40,41], which is consistent with the theory of competitive advantage. The latter affects the geometric size of the non-tradable sector consumption resulting from job growth in the tradable sector; the more diversified the non-tradable sector, the richer and larger the local consumer goods market, leading to more consumer spending by households [31]. The second consists of urban geographical factors. Spatial location, or more precisely the spatial connection among cities, on the one hand, affect labor flow and product trade; on the other hand, the importance of the spatial spillover effect in the multiplier calculation has been confirmed in American cities [19,28]. In the process of rapid urbanization in China, urban production and trade networks are more closely linked, so the spillover caused by this locational relationship is expected to be more significant in China's urban system. In addition to spatial connections, there is also urban size. The influencing mechanism of city size is similar to that of industrial diversity; big cities have a richer variety of consumption and a larger market size. However, because there are already enough living service agencies in big cities, the threshold scale of the multiplier effect will be relatively high.

The first two kinds of factors affect the theoretical expected value of a city's multiplier, and the adjustments of the employment multiplier and job crowding out are the factors that cause the actual multiplier to deviate from the theoretical multiplier. As shown in the previous analysis, the important condition for the emergence of the multiplier is assumed to be the elasticity of labor supply. If the labor supply is inelastic, there will be no multiplier, or the multiplier will be small. Research in Japan has confirmed this view [17]. Housing costs also affect employees' decisions; if the expansion of production in the trading sector leads to a rise in wages and the increase in housing prices is thus greater, there will be a crowd effect on basic employees in the non-tradable sector, reducing the multiplier effect.

3. Materials and Methods

3.1. Research Area

This study selected prefecture-level cities in China as the study area. The choice of spatial scale on the employment problem is very important in the study; prefecture-level cities are the main economic and social units in China. Prefecture-level cities are marked by certain independence and significant levels of difference. Therefore, prefecture-level cities were selected as research subjects.

Choosing Chinese cities as the study area of the employment multiplier reflects unique advantages. First, there are vast differences among many cities in China. The literature focuses on developed countries with high urbanization levels and low labor-supply elasticity. The formation process of China's urban system in the stage of rapid urbanization has unique advantages for studying the formation and evolution of employment problems. Second, in developed countries, particularly the United States, the manufacturing sector is not the main sector of the economy, but that economy is dominated by financial innovation and other activities. Therefore, the impact of changes in the trading sector on non-tradable sectors is not as significant as imagined. China is a manufacturing power, and the migration of manufacturing to developing countries, which is common in developed countries, is rare in China [22]. The driving role of the manufacturing industry in the urban economy is stronger in developing countries than in the case of China, a developed country in a period of transition. Third, both government and market forces affect urban and regional development [42], which is not covered by the current research. Further study of Chinese cities is of great significance for improving the employment multiplier theory and providing reference for developing countries.

China is divided into three regions in the following analysis. The eastern region is mainly coastal provinces with the highest density of population and the most advanced urban economy. It is the most attractive region for internal migrants. The central region is also populous, but urban economy and attractiveness of cities in this region are moderate. The western region is sparsely populated and economically backward. Employment multipliers in cities located in different regions are very likely to be different and their influencing factors are expected to have both similarities and differences as well.

According to the previous analysis, expanding cities were selected as the research unit, so it is necessary to identify and exclude shrinking cities. There are many ways to identify shrinking cities; this study chose the more direct method, which is the reduction of population size, or rate of change. In order to classify the types of urban growth or contraction more carefully, both population growth in municipal districts and population change in municipal districts relative to those of non-municipal districts were investigated. The formula is as follows:

$$\Delta P_s\% = \frac{\left| P_{s(2018)} - P_{s(2008)} \right|}{P_{s(2008)}} - P_a \tag{1}$$

$$\Delta P_{ss}\% = \frac{\left| P_{ns(2018)} - P_{ns(2008)} \right|}{P_{ns(2008)}} - P_a \tag{2}$$

$$D = \Delta P_s\% - \Delta P_{ss}\% \tag{3}$$

where P_s is the municipal population, P_{ns} is the population of non-municipal districts, and D is the difference in the rate of population change between municipal and non-municipal districts. Here, the change in the rate of the population of both municipal and non-municipal districts is subtracted from the national average population growth rate for that period, p_a, to exclude the effects of overall population growth. If D is less than 0, the city is in the contraction stage and was not included in the research scope. As a result, 169 cities were selected for the study (Figure 1).

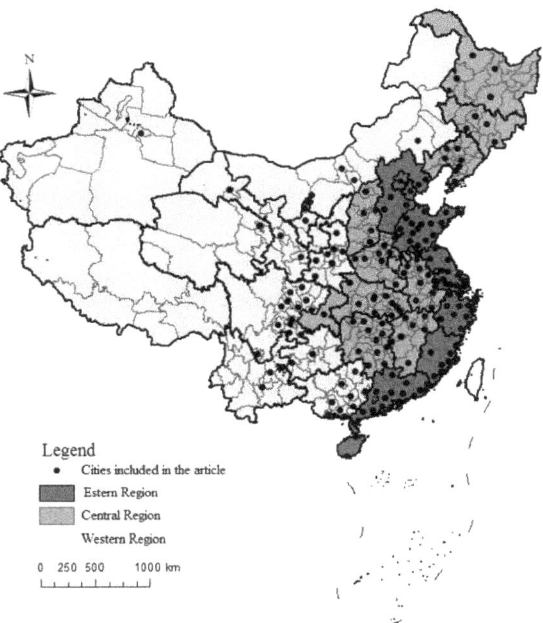

Figure 1. Map of 169 expanding cities included in the article.

3.2. Data

The study used industry statistical data from the China City Statistical Yearbook, from 1998 to 2018, for prefecture-level cities, including manufacturing- and service-sector sectors, which mainly include transportation, warehousing, storage, post, information transmission, computer services and software wholesale, retail accommodation, catering, financial, real estate, leasing and commercial services, scientific research, technical services, geological prospecting, water, environment and public facilities management, education, health, cultural and sports entertainment, resident services, and other services. Industrial structure data use China's economic census data, while location data use Chinese regional urban road network data.

Regarding the number of employees, data from the whole city, rather than those of municipal districts, were selected as the research data for this study because, in the case of China's urban development, the development of central cities is strongly connected with that of the surrounding areas. On the one hand, some manufacturing is concentrated in outlying municipal districts. On the other hand, people who work in municipal districts must live in municipal departments, and their consumption and employment effects will overflow to the municipal districts.

3.3. Methods

3.3.1. Model and Variables

Based on the theoretical analysis and literature review, we developed an empirical model to examine the influencing factors of employment multiplier in Chinese cities. Independent variables include city size, degree of specialization of trade sector, degree of diversification of non-traded sector, locational condition, income, housing cost, and elasticity of labor supply.

$$Mult = \beta_1 CitySize + \beta_2 Div + \beta_3 Prof + \beta_4 Loc + \beta_5 Wage + \beta_6 Hou + \beta_7 Lab + \mu \quad (4)$$

In the model, the urban employment multiplier, represented by *Mult*, is the explained variable, indicating the degree of employment change in non-tradable sectors brought about by the growth of manufacturing jobs in a city. *City size* is the most important factor in the study of urban problems [43,44]. In this study, the permanent urban population was selected as a measure of urban scale. The degree of diversification in the non-tradable sector, represented by *Div*, measures the type and size of the local consumer goods market. The higher the diversification, the higher the local consumer expenditure, as measured by the diversification index. The degree of specialization of the trade sector, represented by *Prof*, affects urban development speed and income level [45], and the specialization index is calculated by using trade-sector employment data. The locational condition, represented by *Loc*, makes up for the industrial and economic relationships between cities, which are neglected in the model [28,46], and uses as a measurement the city's distance to the provincial capital city's expressway. For residents' income levels, we used *Wage*, which is representation in the model, largely to determine their consumption intentions and actual expenditure levels, and cities' average wage levels are used as the measurement index. The cost of living will have a crowding-out effect on employment, especially in low-skilled non-tradable sectors [47]. Housing prices, represented by *Hou*, are a closely watched employment factor in this study [48]. The elasticity of labor supply represented, by *Lab*, is the premise for the realization of the employment multiplier, and the scale of the urban floating population is used as a measure. B_i (i = 1,2,3 …) are coefficients, and μ is a constant term.

3.3.2. Employment Multiplier Calculation Method

On the basis of the model proposed by Moretti [11], we constructed the following model:

$$E^s_{ij,t} - E^s_{ij,t-1} = \beta_0 + \beta_1\left(E^m_{ij,t} - E^m_{ij,t-1}\right) + \alpha City_{i,t-1} + \varepsilon + \mu_\varepsilon \tag{5}$$

where the explained variable is the change in the quantity of employment in the service industry, measured by $E^s_{ij,t} - E^s_{ij,t-1}$, and the core explanatory variable is the change in the employment of the manufacturing industry in $City_i$. β_1 represents the size of the employment multiplier, and *City* refers to other macro variables that may affect both manufacturing and employment changes in a lagged phase. The selection of city-level variables is mainly based on whether they simultaneously influence the changes in the manufacturing and service industries, as well as the degree of influence. This study selects the proportion of government expenditure to GDP, the proportion of fixed-asset investment to GDP, and the ratio of foreign investment to GDP in the current year; μ_ε is the residual, and ε represents the fixed effect controlled in the study.

For the employment of multiplier calculation, the consistency of the OLS estimation error term was not associated with changes in manufacturing employment. This was difficult to meet within the current research context; as a result, instrumental variables were used. However, China's 2003–2015 study showed that the instrumental variable for this period of China's urban employment multiplier in instrumental variable regression results was not large, and the returned benchmark results by the model were robust. The calculation of the instrumental variables was as follows:

$$IV = \sum \frac{Emp^m_{-i,t} - Emp^m_{-i,t-1}}{Emp^m_{-i,t}} \tag{6}$$

where $\sum \frac{Emp^m_{-i,t} - Emp^m_{-i,t-1}}{Emp^m_{-i,t}}$ represents the change in manufacturing employment in regions other than city i. This is the impact of exogenous local economic conditions, which does not affect employment in the local service industry. This study first conducted a fixed-effects regression for the basic model in Equation (1), to obtain the benchmark results and then introduced an instrumental variable to conduct IV-2SLS (Instrumental Variable-Two Stage

Least Squares) regression to measure the multiplier effect of manufacturing on employment creation in the service industry.

3.3.3. Variable Calculation

Industry diversity was calculated by using the Herfindahl–Hirschman Index (HHI) [49], and the calculation formula is as follows:

$$hhi = 1 - \sum_{n=1}^{N_i} S_{i,n}^2 \tag{7}$$

where N_i is industry type in region i, and S_(i,n) is the ratio of the number of persons employed in industry category n in the region to all persons employed in the region. The more evenly the regional employment is distributed across industries, the greater the Herfindahl–Hirschman Index and degree of diversity. When using the proportion of employed persons in various industries to measure the degree of diversity of all urban industries, one must consider the group size of employed people. The ratio of f_i is the number of people employed in the region to the total population of the region.

$$HHI = f_i \cdot \left(1 - \sum_{n=1}^{N_i} S_{i,n}^2 \right) \tag{8}$$

Locational condition was measured by the distance of the city from the expressway of the region's central city (selected as the provincial capital). For the central city itself, the locational condition was calculated by seeing the city as a circle and calculating the radius, using the area from the China Urban Construction Statistical Yearbook.

The regional specialization index is usually used to measure the specialization level of urban industries; that is, the industry with the most employed people in a city is selected as the specialized industry of the city, and the share of employed people in the industry among the total employed people in the city is used as the specialization index. We calculated the following:

$$ZZI_i = max_i\left(S_{ij}\right) \tag{9}$$

where ZZI_i measures the degree of specialization of city *i*. S_{ij} is the industry with the proportionately highest employment share in city *i* of total employment.

The elasticity of labor supply uses the urban floating population as the index, and we calculated the following:

$$Lab = P_a - P_h \tag{10}$$

where *Pa* is the urban resident population from the Economic Statistics Yearbook of Chinese provinces. *Ph* is an urban registered population, for which data are derived from the Statistical Yearbook of Chinese Cities.

The average wage comes from the Economic Statistics Yearbook of Chinese provinces. The housing-price data come from the China housing price quotations website (https://www.creprice.cn/rank/cityforsale.html). Tables 1 and 2 describes the variables selected in this paper, as well as their corresponding indicators, data sources, and descriptive explanations.

Table 1. Index selection, computing, and data sources.

	Index	**Computing and Data Sources**
City size	Urban population	Statistical Yearbook of each province
Diversity	HHI	Formula (8)
Location	Distance to central city	Distance to region's central city
Professionalization	ZZI	Formula (9)
Income	Wage	Economic Statistics Yearbook of each province
Housing costs	Housing price	https://www.creprice.cn/rank/cityforsale.html
Labor supply	Floating population	Formula (10)

Table 2. Descriptive statistics of variables.

	Max	**Min**	**Mean**	**SD**	**Variance**
Multiplier	2.27	0.31	1.67	1.04	14.08
City size	2479.00	30.54	204.02	259.10	6710.99
Diversity	0.85	0.03	0.17	0.09	0.009
Location	1936.00	12.00	227.54	192.10	36,918.90
Professionalization	0.458	0.010	0.027	0.310	1.270
Wage	13,434.00	3817.00	7018.21	4219.07	126,475.11
Housing costs	58,972.00	2638.00	18,256.24	15,127.03	243,657.37
Labor supply	969.17	2.97	128.08	83.94	25,713.14

4. Results

4.1. China's Urban Employment Multiplier

By calculation, the urban employment multiplier in China ranges between 0.31 and 2.27, with a mean of 1.67. This means that introducing enterprises or production activities in a different city leads to the prominence of leading role differences. In areas of high employment multipliers, a trade sector job can produce 2.27 non-tradable sectors of employment growth, and in some areas where the tradable sector shows a poorer stimulating effect, it can only lead to employment in the non-tradable sector. Therefore, the conclusion first affirmed the importance of the trade department to employment growth; some countries are set based on the specific employment policy with regards to this [14]. It turns out that expanding the trade department is an effective means of producing activities to promote employment. However, although the growth of the trade sector has a significantly positive effect on employment, the difference in effect degree in different cities warrants attention. Local governments should not only consider the development of the trade sector but also how to improve the driving effect of the trade sector. The average employment multiplier in cities during the national expansion is 1.67, which is higher than that of other developing countries but close to that of some developed countries [11,17,23]. Studies show that employment multipliers of 1.5–1.7 exist in American cities [20]. This proves the above hypothesis; that is, the employment multipliers of shrinking cities are smaller than those of expanding cities, and the estimated multiplier will be low if the analysis is not differentiated.

In terms of spatial distribution, as shown in Figure 2, the employment multipliers in central and western regions are generally higher than those of coastal cities, which is consistent with other studies [22]. Capital cities in Central China and medium-sized coastal cities, which are China's major manufacturing centers, have higher employment multipliers and have the advantage of productivity. In contrast, the employment multipliers in megacities are not linearly related to the expected size. This is because, when a city is large enough, the non-tradable department within the type and scale of change is less affected by trade department changes, and super megacities are given priority for finance and innovation; so, manufacturing changes brought about by urban employment are less volatile. The employment multipliers in small cities are relatively low; the market sizes and types of consumer goods in small cities are not as large as those in large cities, and the consumer market-pull effect occasioned by the expansion of the trade sector is not as strong as in large cities. In addition, there is a less prominent employment multiplier in Northeast China, which has been related to emigration in recent years. The labor force outflow and the recession of the manufacturing industry make the labor force supply elasticity in Northeast China very low. Meanwhile, the income level in Northeast China also limits the realization of the employment multiplier.

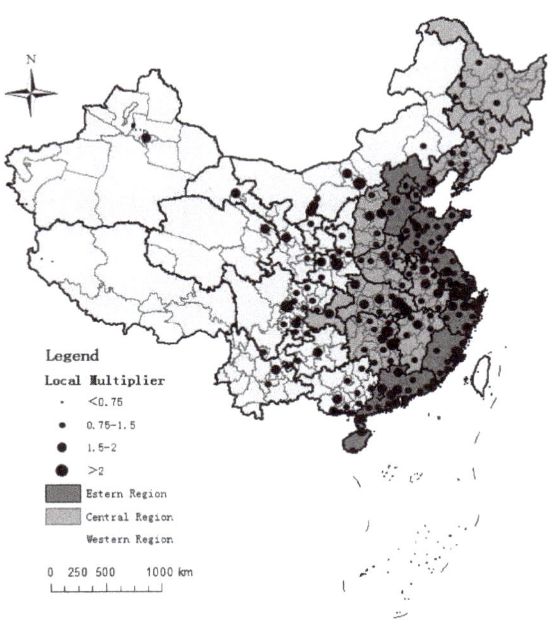

Figure 2. The employment multiplier distribution of expanding cities in China.

After identifying the characteristics of the spatial distribution of employment multiplier, the next obvious question is, how does this pattern connect with city size distribution? In fact, the relationship between city size and employment multiplier has attracted much scholarly attention and varied conclusions have been drawn in different places across the globe [21,27]. In order to have a preliminary understanding of this issue in the case of Chinese cities, we divided the cities into different levels, according to their population sizes, made a statistical description of the employment multiplier of each level of the city, and then made a tentative explanation with the help of the existing literature and theory.

Interesting patterns emerge from the statistics describing the results of the city-scale classification, as can be seen in Table 3. Cities with a population of less than 1 million have lower employment multipliers, with a mean of 0.81, a minimum of 0.31, and a maximum of 1.02. The employment multipliers show a slowly increasing trend with an increase in the city scale because most of these cities are economically underdeveloped within each provincial administrative unit in China. Although the degree of external contact is low, there will theoretically be lower employment spillover and a higher employment multiplier. However, the non-tradable sectors in these cities are small and single in type, which cannot maximize the employment effect brought about by the change in the trade sector. The employment multipliers of cities ranging from 1 million to 3 million are higher than those of the former, with a mean of 1.17, a maximum of 1.54, and a minimum of 0.57. The employment multiplier increased at a faster pace with the expansion of the scale. These cities generally have lower degrees of external contact than regional central cities and have certain levels of economic development and scales of local service industries. Moreover, many of these cities are close to the central city and are affected by the spillover effect of the central city, thus having high employment multipliers. The urban employment multipliers for cities of 3–5 million decrease with the expansion of the urban scale. The employment multiplier reaches the maximum value, with a mean of 1.59, a maximum of 2.27, and a minimum of 1.09. This kind of city is a medium-sized city in China's urban system and is in the stage of rapid growth. Although it is large in scale, it is still at the stage of rapid polarization with a small spillover effect. Moreover, the trade sector based on the manufacturing industry is the main industry in such cities, so the expansion of the trade sector has a great impact on such cities. When the city population is larger than 5 million, the multiplier will start to decline with a mean of 1.08, a maximum of 1.26, and a minimum of 0.94, and the employment multiplier in the megalopolis is then relatively low. Most of these cities are China's regional hubs, with a declining share of manufacturing and a rising share of high-end services, such as financial innovation. Despite a significant employment multiplier, the driving effect of the manufacturing industry is weaker than that of small- and medium-sized cities. At the same time, the central roles of such cities mean that the spillover effect further weakens the employment multiplier.

Table 3. The multipliers of different city size levels.

	Mean	Maximum	Minimum
Less than 1 million	0.81	1.02	0.31
1 million–3 million	1.17	1.54	0.57
3 million–5 million	1.59	2.27	1.09
Larger than 5 million	1.08	1.26	0.94

4.2. Factors Influencing Urban Employment Multipliers in China

As can be seen from the regression results, which are shown in Table 4, there is a significant correlation between the diversification level of the specialized non-tradable sector, average wage and labor-supply elasticity, and the employment multipliers in the urban-scale trade sector nationwide. However, there are obvious differences in the regression between the coastal east and remote west. Therefore, employment policy should not be formulated in accordance with uniform or fixed rules; this also confirms the significance of the study on urban differences.

Table 4. The regression results of influencing factors of employment multipliers.

	Nationwide	Eastern Region	Central Region	Western Region
City size	0.715 *** (0.185)	0.441 ** (0.257)	0.415 ** (0.207)	0.971 ** (0.308)
Diversity	0.169 * (0.026)	0.195 ** (0.021)	0.721	0.031
Location	−0.079 * (0.015)	−0.543 * (0.115)	−0.428	0.153
Professionalization	0.007 *** (0.002)	0.364 ** (0.169)	0.054	0.125
Income	0.281 ** (0.037)	0.034 ** (0.015)	0.513 *** (0.201)	−0.207
Housing costs	0.008	−0.019	0.389	0.009 * (0.003)
Labor supply	0.046 *** (0.021)	0.027 ** (0.003)	0.006 ** (0.003)	0.182
n	169	70	58	41
F	127.394	31.95	27.25	9.929
Adjustment R2	0.757	0.693	0.703	0.455
standard error	0.187	0.297	0.254	0.232

Note: * significance at the 0.05 level, ** significance at the 0.01 level, and *** significance at the 0.001 level. Standard errors are in parentheses.

There is a positive statistical relationship between city size and employment multiplier, nationwide and in the eastern and western regions. This contradicts the description of the full text. As mentioned above, the employment multiplier of a megalopolis tends to decline with the expansion of its scale. The reason for the positive statistical relationship here is that it is mainly medium-sized cities that are expanding in the regression process, and there are fewer megacities. City size has attracted much attention in the study of many urban problems [44,50]. Concepts related to city size include the size and diversification of consumer goods markets into cluster economies and innovative activities [51]. However, a rich variety of consumer goods and large local service requirements are more important reasons for the employment multiplier in big cities than in small cities. City agglomeration and productivity levels mean that local trade department wages and purchasing power are higher than in other cities, which demonstrates an increasing consumer demand for local services. However, as shown in the regional regression results, the multiplier effect of scale is obvious in the western region, because the large cities tend to have greater administrative-level economic power, and the differences between city sizes are actually a reflection of city level and policy bias. In addition, there is a lower degree of external contact in Western China; the employment effect of the trade sector is more easily realized at a local level, and the expansion of the local consumption scale brought about by the expansion of the urban scale is also more obvious. Therefore, to promote urban development and create jobs in the western region, more attention should be paid to the leading role of big cities.

Employment structure is closely related to industrial structure; economic structure and industrial structure decide employment and affect the employment multiplier. As can be seen from the regression results, the degree of tradable sector specialization and the degree of diversification of the non-tradable sector and employment multiplier have positive influences on the employment multiplier. In other words, the more specialized the tradable sector, and the more diversified the non-tradable sector, the higher the urban employment multiplier will be, which is consistent with the theoretical speculation of scholars and empirical research [23,24,32]. However, this rule is not significant in all cities. This pattern is most obvious in the economically developed eastern regions but is not prominent in the central or western regions because of the higher share of agriculture in these regions; this means there is a larger gap between urbanization level and productivity than in the eastern coast. The realization of the employment multiplier is based on the threshold of a certain income level and market size [31], which means that the employment multiplier will be significant when the population size and economic level reach the corresponding levels.

The regression results on wages confirm that income levels are highly correlated with the employment multiplier. The result shows that, the higher the income level of urban residents, the greater the employment multiplier. This correlation is because the multiplier effect in cities is based on consumption [11], and urban residents' incomes are the source of consumer spending; so, the high wage level of the urban employment multiplier is relatively large. The difference between regions shows that, in the central region, the difference in income level is most correlated with the employment multiplier, but this is not significant in the remote western region. Scholars think that the share of residents in local consumer spending is not fixed but changes with the diversity of the local market and scale [31]; there is a larger gap in the size of the market and the types of consumer goods in the western region than in the eastern region.

In previous studies, scholars believed that changes in housing costs would affect the decision-making behavior of job seekers, and if housing costs were high, it would crowd out local services and employment [14]. Thus, in theory, housing costs should present a negative correlation with the employment multiplier, but in Western China, the regression result is positive. The reason is that areas with high housing costs in Western China are those with higher economic and urban scales than in other cities. Although these areas have high housing costs, they show a significantly positive correlation because of their income levels and urban scales.

The elasticity of the labor supply is positively correlated with the employment multiplier, which is consistent with the relevant research conclusions. Labor supply is a relatively important factor in the framework of employment city interpretation proposed by Moretti [11,14], which empirical evidence has also confirmed [17]. Since the employment multiplier is job creation, if a significant local multiplier exists, the city has created more jobs, as well as labor and job matching, which need a large enough labor pool to ensure the realization of an employment multiplier. Moretti believes that if labor-supply elasticity is insufficient, it will cause a non-tradable sector job increase that is not obvious or even decreases with the increase in the trade sector [11].

5. Discussion and Conclusion

This study combines the employment multiplier theory with the theory of urban development stage and selects cities in the process of growth as the research unit to estimate the employment multiplier. It not only has significance for policymaking, but also improves the employment multiplier theory. The change in non-tradable sector employment caused by manufacturing growth in each city is calculated by using the sub-industry data of Chinese prefectural cities, and the difference in this multiplier among different cities is explained by taking city size and trade sector specialization degree as the main explanatory variables. It is of great practical significance to confirm and estimate the differences in employment multipliers between cities. Local differences should be considered in the relevant policies formulated by local governments to promote employment. The multipliers in terms of countries or large regions affirm the importance of changes in the trade sector to regional development, but these are insufficient to answer the question of how to increase the employment multiplier in a city. This study weakened the error range by distinguishing expanding cities from cities which had different multipliers when expanding or shrinking. The multipliers would be underrated if all cities were put into one model.

According to the calculations in this study, the average multiplier of Chinese cities is 1.67, which is close to the previous research results [22]. However, a slightly higher value, not surprisingly, suggests the importance of dividing cities into expanding and shrinking cities. The situation by region and scale is consistent with the research conclusions of other scholars [17,23,32]. It also proves the robustness of the sub-city calculation. Furthermore, it confirms the employment multiplier in the expansion and contraction of the difference between the process and the expansion of the larger number.

In the study of urban problems, urban size once again played an important role. Regression results confirmed the size of the cities in the process of employment change. This is because of the expansion of the city scale and the change in the types of consumer-market scales. The consumption expenditure of urban residents increases with the expansion of the local consumption market. Furthermore, the more significant multilevel employment promotion effect was brought about by the job increase in the trade sector. City size is important not only in the general sense, but also in different regions. It can be seen from the regression results that the coefficient of urban size in the western region is larger than that in other regions, indicating that the efficiency advantage of large cities should be brought into play in the western region with sparse population; this is also revealed laterally, so small cities have greater difficulty in realizing the same multiplier effects. The specialization of the tradable sector is the guarantee of employment multipliers and even urban development. This is reflected in employment level, and urban residents' incomes have far-reaching influences on the speed and sustainability of urban development; this is consistent with Michael Stoper's conclusion [41]. Therefore, the ability to produce tradable products professionally warrants close attention.

China differs from developed countries regarding the impact of factors such as labor costs and housing costs on the employment multiplier. In Moretti's explanation of the multiplier, factors such as labor cost and housing cost have a correction and adjustment effect on the employment multiplier [11,14]. The results show that these factors are not significant in China, which is also one of the differences between developing countries and developed countries. The low level of urbanization in developing countries means that many agricultural workers can enter the labor pool in cities, thus improving the elasticity of the urban labor supply. This weakens the crowding-out effect of housing costs and other factors, even showing a positive correlation between housing costs and multipliers in some regions. We should consider learning from the experiences of developed countries in the process of regional development. The differences between China and developed countries in terms of institutional urbanization level, labor force structure, urban system, industrial structure, and other aspects mean that the influencing mechanism of China's urban employment problem is different from that of developed countries.

According to the research conclusions of this study, the following policy recommendations are proposed. The difference in the employment multiplier determines the difference in employment promotion policy, which should be considered when making this policy. Specifically, in the central and western regions, the trade sector is the main driving force for employment. Therefore, in the central and western regions, efforts should be made to develop the scale of the trade sector and promote the expansion of the urban scale. In large cities and eastern regions, the employment effect of the trade sector is still significant, but the degree of effect is not as great as that of small- and medium-sized cities. Therefore, the degree of specialization of the trade sector and diversification of the service industry should be improved to maximize the employment effect of the trade sector in the urban consumption scale.

In this study, time-series data were used to calculate the employment multiplier of a single city, which may have led to different degrees of error because the contribution of spatial variables differs between different cities. Therefore, the model needs to be further improved. As an extension and sublimation of this study, the other study uses panel data to explore the periodic changes in employment multipliers in cities, according to their development stages, also analyzing the mechanisms of employment multipliers in shrinking cities [45].

Author Contributions: Conceptualization, D.H. and T.L.; methodology, D.H., H.H., and T.L.; software, H.H.; validation, D.H. and H.H.; formal analysis, H.H.; resources, D.H. and T.L.; data curation, D.H. and H.H.; writing—original draft preparation, D.H. and H.H.; writing—review and editing, D.H. and T.L.; visualization, H.H.; supervision, D.H. and T.L.; project administration, D.H.; funding acquisition, D.H. and T.L. All authors have read and agreed to the published version of the manuscript.

Funding: This research was funded by the National Natural Science Foundation of China (41801146, 41741009), the Humanities and Social Science Youth Foundation of the Ministry of Education of China (18YJC840022), and UKRI's Global Challenge Research Fund (ES/P011055/1).

Data Availability Statement: Data used in this study are publicly available from https://tongji. oversea.cnki.net/chn/navi/HomePage.aspx?id=N2018050234&name=YZGCA&floor=1.

Conflicts of Interest: The authors declare no conflict of interest.

References

1. Mortensen, D.T.; Pissarides, C.A. Job creation and job destruction in the theory of unemployment. *Rev. Econ. Stud.* **1994**, *61*, 397–415. [CrossRef]
2. Essletzbichler, J. The Geography of Job Creation and Destruction in the U.S. Manufacturing Sector, 1967–1997. *Ann. Assoc. Am. Geogr.* **2004**, *94*, 602–619. [CrossRef]
3. Wilson, W.J. *When Work Disappears: The World of the New Urban Poor*; Alfred Knopf: New York, NY, USA, 1996; Volume 21.
4. Coffey, W.J.; Shearmur, R. Factors and correlates of employment growth in the canadian urban system, 1971–1991. *Growth Chang.* **1998**, *29*, 44–66. [CrossRef]
5. Blien, U.; Suedekum, J.; Wolf, K. Local employment growth in West Germany: A dynamic panel approach. *Labour Econ.* **2006**, *13*, 445–458. [CrossRef]
6. Cai, F.; Chan, K.W. The global economic crisis and unemployment in China. *Eurasian Geogr. Econ.* **2009**, *50*, 513–531. [CrossRef]
7. Kannappan, S. Urban employment and the labor market in developing nations. *Econ. Dev. Cult. Chang.* **1985**, *33*, 699–730. [CrossRef]
8. Xu, W.; Pan, Z.; Wang, G. Market transition, labor market dynamics and reconfiguration of earning determinants structure in urban China. *Cities* **2018**, *79*, 113–123. [CrossRef]
9. Lewis, W.A. Economic development with unlimited suoolied of labor. *Manch. Sch.* **1954**, *22*, 139–191. [CrossRef]
10. McGregor, P.; McVittie, E.P.; Swales, J.K.; Yin, Y.P. The neoclassical economic base multiplier. *J. Reg. Sci.* **2000**, *40*, 1–31. [CrossRef]
11. Moretti, E. Local multipliers. *Am. Econ. Rev.* **2010**, *100*, 373–377. [CrossRef]
12. Storper, M. Why do regions develop and change? The challenge for geography and economics. *J. Econ. Geogr.* **2011**, *11*, 333–346. [CrossRef]
13. Sunde, T. Foreign direct investment, exports and economic growth: ADRL and causality analysis for South Africa. *Res. Int. Bus. Finance* **2017**, *41*, 434–444. [CrossRef]

14. Moretti, E.; Thulin, P. Local multipliers and human capital in the United States and Sweden. *Ind. Corp. Chang.* **2013**, *22*, 339–362. [CrossRef]

15. Keynes, J.M. The general theory of employment. *Q. J. Econ.* **1937**, *51*, 209–223. [CrossRef]

16. Mathur, V.K.; Rosen, H.S. Regional employment multiplier: A new approach. *Land Econ.* **1974**, *50*, 93–96. [CrossRef]

17. Kazekami, S. Local multipliers, mobility, and agglomeration economies. *Ind. Relat. A J. Econ. Soc.* **2017**, *56*, 489–513. [CrossRef]

18. Moritz, T.; Ejdemo, T.; Söderholm, P.; Wårell, L. The local employment impacts of mining: An econometric analysis of job multipliers in northern Sweden. *Miner. Econ.* **2017**, *30*, 53–65. [CrossRef]

19. Gerolimetto, M.; Magrini, S. A spatial analysis of employment multipliers in the US. *Lett. Spat. Resour. Sci.* **2016**, *9*, 277–285. [CrossRef]

20. van Dijk, J.J. Local employment multipliers in US cities. *J. Econ. Geogr.* **2017**, *17*, 465–487.

21. Osei, M.J.; Sengupta, S. Heterogeneity in the local employment multipliers in the United States. *Growth Chang.* **2019**, *50*, 880–893. [CrossRef]

22. Wang, T.; Chanda, A. Manufacturing growth and local employment multipliers in China. *J. Comp. Econ.* **2018**, *46*, 515–543. [CrossRef]

23. Jones, M.D.; Yang, L. Regional job multipliers. *Appl. Econ. Lett.* **2018**, *25*, 1342–1345. [CrossRef]

24. Goos, M.; Konings, J.; Vandeweyer, M. Local high-tech job multipliers in Europe. *Ind. Corp. Chang.* **2018**, *27*, 639–655. [CrossRef]

25. Diacon, P.E.; Maha, L.G. The relationship between income, consumption and GDP: A time series, cross-country analysis. In *Procedia Economics and Finance*; Iacob, A.I., Ed.; Elsevier Science BV: Amsterdam, The Netherlands, 2015; Volume 23, pp. 1535–1543.

26. Garcia, C.Q.; Velasco, C.A.B. Searching for complementary technological knowledge and downstream competences: Clustering and cooperation. *Int. J. Technol. Manag.* **2006**, *35*, 262–283. [CrossRef]

27. Mulligan, G.F.; Vias, A.C. Place-specific economic base multipliers. *Environ. Plan. B Plan. Des.* **2011**, *38*, 995–1011. [CrossRef]

28. Anselin, L. Spatial externalities, spatial multipliers, and spatial econometrics. *Int. Reg. Sci. Rev.* **2003**, *26*, 153–166. [CrossRef]

29. Li, G.; Li, F. Urban sprawl in China: Differences and socioeconomic drivers. *Sci. Total. Environ.* **2019**, *673*, 367–377. [CrossRef]

30. Kline, P.; Moretti, E. People, places, and public policy: Some simple welfare economics of local economic development programs. In *Annual Review of Economics*; Arrow, K.J., Bresnahan, T.F., Eds.; Annual Reviews: Palo Alto, CA, USA, 2014; Volume 6, pp. 629–662.

31. Pred, A.R. *The Spatial Dynamics of U.S. Urban-Industrial Growth*; MIT Press: Cambridge, MA, USA, 1966; Volume 22, p. 171.

32. Fleming, D.A.; Measham, T.G. Local job multipliers of mining. *Resour. Policy* **2014**, *41*, 9–15. [CrossRef]

33. Fujita, M.; Krugman, P.; Venables, A.J. *The Spatial Economy: Cities, Regions, and International Trade*; MIT Press, Cambridge/Mass: London, UK, 2001.

34. Robinson, J.V. *Introduction to the Theory of Employment*; Macmillan: London, UK, 1937.

35. Lamo, A.; Pérez, J.J.; Schuknecht, L. The cyclicality of consumption, wages and employment of the public sector in the euro area. *Appl. Econ.* **2013**, *45*, 1551–1569. [CrossRef]

36. Bishop, P. Spatial spillovers and employment growth in the service sector. *Serv. Ind. J.* **2009**, *29*, 791–803. [CrossRef]

37. Edwards, E.O. *Employment in Developing Nations*; Columbia University Press: New York, NY, USA, 1974.

38. Berry, B.J.L. City size distributions and economic development. *Econ. Dev. Cult. Chang.* **1961**, *9*, 573–588. [CrossRef]

39. Batten, D.F. Network cities: Creative urban agglomerations for the 21st century. *Urban. Stud.* **1995**, *32*, 313–327. [CrossRef]

40. Plumper, T.; Graff, M. Export specialization and economic growth. *Rev. Int. Polit. Econ.* **2001**, *8*, 661–688. [CrossRef]

41. Kemeny, T.; Storper, M. Is specialization good for regional economic development? *Reg. Stud.* **2015**, *49*, 1003–1018. [CrossRef]

42. Farrell, K.; Westlund, H. China's rapid urban ascent: An examination into the components of urban growth. *Asian Geogr.* **2018**, *35*, 85–106. [CrossRef]

43. Gabler, L.R. Population size as a determinant of city expenditures and employment: Some further evidence. *Land Econ.* **1971**, *47*, 130–138. [CrossRef]

44. Parkinson, M.; Meegan, R.; Karecha, J. City size and economic performance: Is bigger better, small more beautiful or middling marvellous? *Eur. Plan. Stud.* **2014**, *23*, 1054–1068. [CrossRef]

45. Abdel-Rahman, H.M. When do cities specialize in production? *Reg. Sci. Urban. Econ.* **1996**, *26*, 1–22. [CrossRef]

46. Krumm, R.; Strotmann, H. The impact of regional location factors on job creation, job destruction and employment growth in manufacturing. *Jahrb. Reg.* **2012**, *33*, 23–48. [CrossRef]

47. Ernst, E.; Saliba, F. Are house prices responsible for unemployment persistence? *Open Econ. Rev.* **2018**, *29*, 795–833. [CrossRef]

48. Saks, R.E. Job creation and housing construction: Constraints on metropolitan area employment growth. *J. Urban. Econ.* **2008**, *64*, 178–195. [CrossRef]

49. Matsumoto, A.; Merlone, U.; Szidarovszky, F. Some notes on applying the Herfindahl–Hirschman Index. *Appl. Econ. Lett.* **2012**, *19*, 181–184. [CrossRef]

50. Zhang, W.; Yang, D.; Huo, J. Studies of the relationship between city size and urban benefits in China based on a panel data model. *Sustainability* **2016**, *8*, 554. [CrossRef]

51. Abdel-Rahman, H.M. Agglomeration economies, types, and sizes of cities. *J. Urban. Econ.* **1990**, *27*, 25–45. [CrossRef]

Article

A Multidimensional Evaluation Approach for the Natural Parks Design

Vincenzo Del Giudice [1], Pierfrancesco De Paola [1,*], Pierluigi Morano [2], Francesco Tajani [3] and Francesco Paolo Del Giudice [1]

1   Department of Industrial Engineering, University of Naples "Federico II", Piazzale Vincenzo Tecchio 80, 80125 Naples, Italy; vincenzo.delgiudice@unina.it (V.D.G.); francesco.delgiudice@libero.it (F.P.D.G.)

2   Department of Science of Civil Engineering and Architecture, Polytechnic of Bari, Via Orabona 4, 70125 Bari, Italy; pierluigi.morano@poliba.it

3   Department of Architecture and Design, University of Rome "Sapienza", Via Flaminia 359, 00196 Rome, Italy; francesco.tajani@uniroma1.it

*   Correspondence: pierfrancesco.depaola@unina.it

Abstract: The design of a natural park is generated by the need to protect and organize, for conservation and/or for balanced growth, parts of the territory that are of particular interest for the quality of the natural and historical–cultural heritage. The necessary tool to support the decision-making process in the design of a natural park are the financial and economic evaluations, which intervene in three successive steps: in the definition of protection and enhancement levels of the park areas; in the choice of the interventions to be implemented for the realization of these levels of protection and enhancement; in determining and verifying the economic and financial results obtainable from the project execution. This contribution deals with aspects and issues relating to the economic and financial evaluation of natural park projects. In particular, an application of the "Complex Social Value" to a concrete case of environmental design is developed on the basis of the elements that can be deduced from a feasibility study of a natural park: the levels of protection and enhancement of the homogeneous areas of the natural park are preliminarily defined, and the choice of the design alternative to be implemented is, therefore, rationalized with multicriteria analysis.

Keywords: complex social value; natural park design; environmental design; multicriteria analysis

Citation: Del Giudice, V.; De Paola, P.; Morano, P.; Tajani, F.; Del Giudice, F.P. A Multidimensional Evaluation Approach for the Natural Parks Design. *Appl. Sci.* **2021**, *11*, 1767. https://doi.org/10.3390/app11041767

Academic Editor: Elmira Jamei

Received: 29 January 2021
Accepted: 12 February 2021
Published: 17 February 2021

1. Introduction

The design of a natural park consists in the conception of a territorial system (park system) whose compositional and structural characteristics derive from a careful territory examination and its subdivision into homogeneous areas that constitute the park "subsystems". Within the individual areas, the activities are established on the basis of a summary judgment between the vulnerability degree and the social utility degree of the existing resources and emergencies.

The determination of the vulnerability degree involves multidisciplinary skills, which obviously depend on the nature and characteristics of the resources under study. The determination of the social utility degree is a specifically evaluative problem and can be summed up in the value that the community attributes to resources.

Since these are goods of a purely qualitative nature, not reproducible, belonging to the community and, therefore, by definition not exchangeable and capable of presenting values independent of use, the criterion of "complex social value" can be used for their valuation, corresponding to the "total economic value", grouping in its composition the preferences of all the subjects directly and indirectly involved in the formulation of the value judgment [1].

The complex social value synthesizes both economic needs and those that cannot be connected with objectives of pure efficiency. Therefore, it carries out a cognitive function aimed at revealing the multiple social expectations regarding increasingly scarce natural resources, whose use program, in the processes of territorial redevelopment or development, is aimed at a balanced qualitative–quantitative growth of the socio-economic and ecological–environmental components [2,3].

The search for the optimal degree of integration between various modes of growth cannot be separated from a composite evaluation set that can be formulated with a view to subjecting each sub-system to types of growth selected on "merit values" attributed by the community to existing resources.

Generally, it can be accepted that activities compatible with the objectives of exclusive protection must be localized in the subsystems that have a "high" complex social value. Any exceptions regarding the possibility of providing moderate forms of transformation of the environmental components can only be considered following a judgment of compatibility between the impacts originating from the transformation and the qualitative characteristics of the resources. Similarly, mixed objectives of protection and enhancement and objectives of mere enhancement may be pursued in areas that have a complex social value, respectively, "medium" and "low".

However, the classification of differentiated levels of value can only be carried out after having explained the various aspects of the complex social value of the resources and emergencies present in the park areas. This value, as is known, is given by the sum of two components: the "use value" and the "use-independent value". The use value is connected to the use of a certain resource and arises from the flow of collective utility consequent to the use, even indirect and current (vicarious value) or future (option value, bequest value), of the resource itself. The value independent of use, in turn, is represented by the so-called existence value, which depends solely on the fact that the resource "exists", regardless of its use/enjoyment of a direct and indirect type [1–4].

To express in economic terms the use value and the use-independent value of a resource, it is possible to consider various valuation methods.

In particular, the value associated with the direct use of a natural resource can be derived from a demand curve constructed through measures of willingness to pay or accept (direct methods), or by using "proxy" variables of value (indirect methods). The "willingness to pay" can also be used to express in economic terms the indirect use value and the rate of the complex social value independent by resource use. The valuation procedure generally applied is the Contingent Valuation (CV), which makes it possible to prefigure a hypothetical market for the asset, from which the value is deducted. The evaluation is thus carried out directly, without resorting to parameters that act as a proxy for the unknown value [5].

However, with economic evaluations only, it is not possible to express the various aspects of the complex social value of a resource. This is because some qualitative environmental components (i.e. the landscape, aesthetic, cultural or ecological) generally escape a monetary representation. It, therefore, becomes necessary, in order to correctly estimate the complex social value, to develop a disaggregated qualitative–quantitative evaluation with respect to a certain number of criteria.

The significance of the complex social value will obviously depend on the congruity of the individual assessments and the ability to compose them according to a multidimensional profile.

In order to define the protection levels and enhancement to be achieved in the park areas, it will then be necessary to identify a priority ranking among the sub-systems that takes into account the complex social value of each one. In this way, it will be possible to compare the different values attributed to the resources and emergencies present in the individual sub-systems, and subsequently to establish the methods of environmental protection to be implemented [2–4].

In the framework outlined, it must be said that among the major weaknesses of sustainable development is that it is not always possible to adequately measure the level of sustainability achieved by a particular activity or government/institution. There is a lack of knowledge on which environmental issues should be incorporated into the economic calculation and on how sustainability can be measured.

Sustainability is a multidimensional concept: the economic, social and environmental aspects must be considered simultaneously. This can be adequately considered through the complex social value, which is expressed through a set of multidimensional indicators. For this reason, the research question of the study consists of an attempt to combine the aspects of sustainability with assessments from the point of view of the community (economic, social, environmental), putting them in relation to each other.

2. Literature Review

With reference to decision-making processes, valuation systems can assume different meanings especially if they are related to spatial planning.

The issues with value in planning were examined by Campbell [6], who analyzed how planners can make ethical or qualitative judgements based on a critical understanding of the decision context considered.

Planning issues require evaluation methods based on complex value-focused thinking: this helps to articulate values, identify decision opportunities and create alternatives [7].

The "complex social value" of a context and its resources was considered by Fusco Girard [8]. Further, this value expresses a system of immaterial relations, its specific character and identity [9].

Another concept of value complex was formulated by Zeleny [10], conceiving it as a metacriterion: an expression of a cognitive equilibrium integrated and rooted in specific contexts [11].

Complex systems can reflect only a specific subset of possible representations [12,13]; thus, the public-decision problems must be used to choose a definition of "value" under an operational profile, although different policy goals may specify different aspects or definitions of value. Additionally, multiple values correspond to as many multiple forms of knowledge [14,15].

With regard to the importance of "social" decision making, the ways in which values, preferences and alternative knowledge are derived from interactions with the social environment, Larner and Le Heron highlighted the context of the decision-making environment, both in spatial and scalar terms [16].

Decisions based on complex values enable a better focus on the decision problem structure [17], where complex issues can also be complicated, not structured, difficult to manage or ambiguous [18–21].

Again, complex values are connected to the context and the decision framework, and they take shape through physical, environmental, social and economic environments [22].

With regard to the evaluation in planning, Alexander [23] focuses on the concept of planning-evaluation proposed by Lichfield [24]: evaluation is conceived as closely embedded in planning, evolving with it. Then, evaluation method evolution reflects the planning process interaction with the diversity and complexity of knowledge, favoring new approaches and methods focused on complex multimethod evaluation systems [23,25,26].

Among the applications of complex social value for the enhancement of ecosystems, the most representative studies are those of Sherrouse et Al. [27], Fulgencio [28] and Fagerholm et Al. [29]. In particular, Sherrouse et Al. [27] developed a tool to assess, map and quantify nonmarket values perceived by various groups of ecosystem stakeholders; this has two main objectives: evaluate how effectively the value index developed reproduces results from more common statistical methods of social-survey data analysis; examine how the spatial results provide additional information that could be used by stakeholders to better understand more complex relationships among stakeholder values, attitudes and preferences. Fulgencio [28] tries to clarify the understanding of social value in an innovation ecosystem, as a tool to aid science park orchestrators or managers to manage the expectations of social and nonsocial actors. Fagerholm et al. [29] synthesized the existing analysis methods applied to the data collected through participatory mapping approaches, with the aim to guide both novice and experienced practitioners in the field of participatory mapping.

However, in the examined studies, high attention should be paid to the fact that an assessment based solely on economic or social impacts does not always guarantee a fair integration of multidimensional values in the decision-making process, because it does not it take into account the many temporal phases of spatial context transformation.

3. Materials and Methods

The process of determining and evaluating strategic choices stimulates the search for increasingly objective systems and/or selection criteria that are not influenced by endogenous factors. This problem is particularly relevant if investment projects need to have public funding.

The constraints deriving from economic, social and environmental issues often contrast with the design needs, making choices influenced by value judgments indispensable. It is precisely in the need to make choices and in the opportunity to support them, even scientifically, that statistical tools are inserted, including multicriteria analysis.

Despite the great variety of multidimensional evaluation methods, they all have two elements in common: the existence of multiple evaluation criteria, often conflicting, for which there are different units of measurement; the possibility of a multidisciplinary approach [3,4].

A classification of multidimensional methods enables their subdivision into discrete multicriteria methods and continuous multiobjective methods.

Continuous methods can include an infinite number of choice possibilities (they concern the identification of the best choice within an infinite set of alternatives, given the pre-established constraints), while discrete methods take into consideration a finite and explicit number of feasible decision alternatives (actions, plans, interventions or projects that are alternative to each other). The latter, therefore, is better suited to be used downstream of evaluations when it is a question of comparing a finite number of opposing "alternatives".

The variety of tools offered by Multicrier Analysis includes techniques regulated by simple algorithms (dominance analysis, for example) and techniques that use more complex algorithms, among which the most frequently used are the Concordance Analysis and the Analytical Hierarchy Process (AHP). Other methods are the Electre method, the Evamix method and the Topsis method [30–39].

Giaoutzi and Nijkamp [35] gave, through an equilateral triangle diagram, a definition of sustainable development in which three dimensions are combined: economic, environmental and social. According to this triangle, sustainable development can be seen as a combination of the position of the economist, the opinion of the sociologist and the attitude of the environmentalist. Making choices will, therefore, mean recognizing and accepting priorities and through them favoring one position over another (establishing criteria).

In application practice, among the most used multicriteria evaluation procedures is the qualitative multicriteria analysis developed by Nijkamp [36–39], which is very useful, especially in the presence of little information on the effects of projects. This procedure consists in identifying classes of importance and effectiveness, then assigning preference scores and calculating how many times a given design alternative falls into a certain importance/effectiveness class. On the basis of the index found, a table of combined frequencies is constructed, in which each element indicates how many times a design alternative proves to be more or less effective and important. Although considered particularly easy to use, the Nijkamp method has the limit of establishing whether one project is better than another, but not to what extent, like any other method of qualitative evaluation.

4. Application of the Complex Social Value: Research Steps

After a preliminary overview of the territorial context of interest, the research phases can be summarized as follows:

1. Subdivision of the park area into homogeneous territorial zones (sub-systems) for morphological, utilization and anthropization characteristics;
2. Classification of homogeneous areas according to their complex social value;
3. Definition of the activities to be started for the constitution of the natural park on the basis of the classification referred to in point 1;
4. Identification of design alternatives;
5. Determination of the preferability order for the design alternatives.

4.1. Territorial Context

The system of the Picentini Mountains extends from the province of Avellino to that of Salerno, in Campania. It is bordered to the west by the Irno river valley, to the east by the Alto Sele valley, to the south by the plain of Battipaglia and to the north by the Ofanto river and the route of the ancient Via Appia. It is, therefore, placed between the "Neapolitan conurbation"—which is a dense urban and semi-urban agglomeration that extends continuously on the coastal strip between Cuma and the west of Naples, and Eboli and the south of Salerno—and the inland areas of Alta Irpinia. The Picentini Mountains include, in a landscape continuum of particular environmental interest, a set of reliefs and valley bottoms with evident and accentuated characteristics of morphological and landscape unity. The peaks of Monte Mai, Polveracchio, Calvello and Accelica are crowned by the higher reliefs, Mount Terminio (1783 m) and Mount Cervialto (1809 m). The system is rich in tall forests and spring waters, which give rise to the Sele, Ofanto, Calore, Sabato, Picentino and Tusciano rivers. The waters are partly used by hydroelectric plants and partly destined for drinking purposes in the Campania and Puglia regions. Of the entire system, the park area extends over approximately 14,000 hectares and is roughly delimited: to the east, by the administrative border of the "Valle dell'Irno" mountain community, coinciding with the ridge limit that determines the natural division and structural of the territory on two sides (western and eastern); to the north, west and partly to the south by the Salerno-Avellino highway and railway, which mark the border strip characterized by strong anthropization; to the south, for the portion not delimited by the highway, by the line of the road connecting the smaller towns. The land surrounding the inhabited centers is covered by vineyards, chestnut groves and mostly tree-lined arable land, managed in the direct economy by the farmers. The zootechnical activity, made up of sheep, goat and cattle breeding, is fragmented into small family farms and is constantly shrinking. The industrial initiatives are mainly located along the southern and western axes of the area, with tanneries in Solofra, spinning mills and small foundries in the valley areas. A good source of income is given by the production of wood and the small industries connected to it.

4.2. Homogeneous Territorial Zones

For the case study, the information relating to a feasibility study prepared for the enhancement of the park was used as a reference. In this reference, the park area is already divided into homogeneous zones. The subdivision was made on the elements collected with the specialist investigations carried out on the main environmental components of the park area:

- physical environment;
- Flora–vegetational–forest environment;
- Wildlife environment;
- Historical–cultural and anthropic environment.

The homogeneous areas identified have the following denominations and characteristics:

- Zone A—Area of natural environment. It includes the cacuminal belt of the mountains above the chestnut area and the areas of difficult access between mounts, where environmental resources are in almost optimal conditions.
- Zone B—Area of semi-natural environment. It includes the influence the area of a water basin. This area constitutes a defined and limited ecosystem, and it is characterized by an environmental balance determined by careful resource use. It falls within

the belt of the mainly western mountain slope, of pre-eminent landscape interest due to the scenery effect it produces on the inhabited centers located along the foothills.

- Zone C—Area of agro-forestry and agricultural environment. It includes the flat areas that extend around the inhabited centers, as well as the foothills and the valley floors, which, due to orographic characteristics, allow agricultural land use.
- Zone D—Area of urban environment. It includes the inhabited centers falling within the natural park perimeter.

The definition of these zones is consistent with the indications provided by European Community for the harmonization, at European level, of the zoning system for protected areas.

The percentage distribution of the 14,000 hectares that make up the park area among the homogeneous areas is identified as follows:

- Zone A: 12.70%;
- Zone B: 32.40%
- Zone C: 33.60%;
- Zone D: 21.30%.

Tables 1–4 show the summary data relating to the environmental heritage and production activities present in the study area.

Table 1. Naturalistic and anthropic emergencies.

Zone	Location	Naturalistic Emergencies	Monumental–Artistic Emergencies
A	Calvanico		Church of San Michele
	Pizzo San Michele	Presence of fossils	
	Serre of Torrione		
	Mount Monna		-
	Faggeto Valley	Beech forest, limestone ridges of Pizzo San Michele (1576 mt.) and Mount Mai (1607 mt.)	
B	Candelito Valley		Paleolithic finds 40,000 A.C.
	Fisciano	-	
	Prepezzano	Eastern edge of tectonic faults	Convent of San Michele
	Angel Cave	Natural caves	
	Bad Cave		-
C	Macchione		Villa Rustica II sec. A.C., San Cipriano Picentino Castle, Medieval settlements
	Mount Vetrano	-	
	Campo di Valle		-
	Castiglione dei Genovesi	Presence of fossils	Villa I sec. D.C.
	Mount Tubenna		Convent
	Pezzano	-	Mount Tubenna
D	Passatoia		Italic tombs V sec. A.C.
	Pozzillo		Imperial Villa II sec. D.C.
	Calvanico		Churh of San Salvatore XVII sec.
	Prepezzano	-	Cathedral with San Nigia Tower
	Fisciano		Convent of Capuchins
	Solofra		Collegiata of San Michele, Baroque cathedral XVII sec., Castle of Rota, Old tanneries of the early industrial period

Table 2. Flora and wildlife present in the park area.

Zone	Main Flora–Vegetational–Forest Presences	Main Wildlife			
		Mammals	Amphibians	Reptiles	Birds
A	Beech	Marten			Pregrine Falcon, Landrio
	Chestnut	Weasel	-		Buzzard
	Holm oak	Badger			Hoopoe
	Aquilina fern	Wolf			Cotressola
B	Oak	Surnottolo		Orbettina	Codirosso
	Cerro		-	Biacco	Frosone
	Chestnut	Flour			Sorpone
C	Orniello		Tree frog	Cervone	Taccola
	Hornbeam	-	Fire salamander	Grass snake of the collar	Zigolo
D	-	-	-	-	-

Table 3. Land use and productive activities.

Zone	Land Use	Productive Activities
A	Prevalence of woods, chestnut and beech woods, grazing and bushy pasture	Agriculture
B	Woods, chestnuts, hazelnuts and beeches, with the presence of specialized crops: citrus and olive groves; grazing and bushy pasture	
C	Prevalence of arable land and specialized crops: citrus and olive groves; grazing and bushy pasture	Agriculture and farming
D	Prevalence of arable land and specialized crops: citrus and olive groves	Agriculture and farming, industrial activities: light manufacturing and tanneries

Table 4. Land use and productive activities.

Zone	Location	Mining Activities	Landfills
A	Calvanico	Fossils	Illegal landfills
	Costa Grande	-	Landfill of municipal solid waste
B	-	-	-
C	Rocca dell'Aquila	Disused quarry	-
	Fisciano	-	Landfill of municipal solid waste
D	San Cipriano Picentino	Active quarry	-

4.3. Protection Levels: Classification of Homogeneous Areas

The complex social value enables the measure of the protection degree to be achieved in the individual park areas. It must, therefore, be determined for each homogeneous zone. The comparison of the results leads to the ranking and classification of the zones: zones with the highest complex social value are those in which it is preferable not to carry out any transformation; for areas with a lower complex social value, modifications of the use characters may be envisaged.

Multicriteria qualitative–quantitative analysis was applied to estimate the complex social value for the single homogeneous zones.

Based on the reference literature [30–45], the criteria considered are the following:

- C1. landscape/perceptual quality;
- C2. archaeological/cultural quality;
- C3. vegetation quality;
- C4. wildlife quality;
- C5. quality of production activities;
- C6. accessibility.

Criteria C1 to C4 reflect the "intrinsic" quality of the area (value independent of use); C5 and C6 reflect the "extrinsic" quality (use value). The indicators selected to express the chosen criteria are:

- I1. level of presence in the area of sites with particular aesthetic value;
- I.2. state of archaeological/cultural emergencies;
- I.3. level of presence of flora–vegetational species with particular interest and rarity;
- I.4. level of presence of rare wildlife species;
- I.5. average annual income produced in the area;
- I.6. average market price of land with the same destination than that prevailing in the area.

The assessments corresponding to indicators I1 to I4 are based on an ordinal scale. The assessments relating to I5 and I6, as they are of an economic type, are based on a cardinal scale.

The assessment summary, with respect to the selected criteria, is shown in Table 5, whose rows explain the complex social values for the four homogeneous areas between which the study area was divided.

Table 5. Summary of complex social value with respect to the selected criteria.

Zone	C1	C2	C3	C4	C5	C6
A	4	2	3	4	2	1
B	3	2	3	3	2	2
C	2	3	2	2	3	2
D	1	4	1	1	4	3

The value judgments contained in Table 5 are expressed through ordinal numbers ranging from 1 to 4, with 1 and 4 equal to the minimum and maximum values, respectively. The attribution of the values corresponding to the criteria from C2 to C6 did not cause difficulties, as it was possible to obtain the results of specialist surveys on the main environmental components of the park area. The definition of the value relating to criterion C1 was more complicated due to the scarce information available. However, this latest information has been integrated with data from a survey carried out through interviews.

Additionally, economic assessments are expressed in ordinal scale, since the multicriteria analysis purpose is to define an ordering of the areas, a ranking according to their complex social value. To define the priority order of the homogeneous zones, it is necessary to assign a "weight" to each of the considered evaluation criteria. Each combination of weights corresponds to a different sorting of the zones. It is, therefore, a question of identifying an overall ranking that takes into account all the possible weighting systems of the criteria.

The overall ordering of the park areas for the possible weight combinations was obtained with the regime analysis developed by Nijkamp and Hinloopen [36,37]. In fact, the regime analysis makes it possible to determine the overall priority of the park areas even if only ordinary information is available.

Table 6 shows the zones ordering according to the attributions (w_k) of different weights to the six evaluation criteria considered. The weights are expressed using ordinal numeric symbols.

Table 6. Zones ordered according to the weight attributions.

Weights						Zone			
w1	**w2**	**w3**	**w4**	**w5**	**w6**	**A**	**B**	**C**	**D**
1	2	3	4	5	6	4	3	1	2
6	1	2	3	4	5	4	3	1	2
5	6	1	2	3	4	4	3	1	2
4	5	6	1	2	3	4	2	1	3
3	4	5	6	1	2	4	3	1	2
2	3	4	5	6	1	3	2	1	4

Table 6 highlights that zone A was the one that presented, in almost all cases, the highest preferability and, therefore, the highest complex social value. The preference for zone B was slightly lower, then zone D followed and, finally, zone C, which in all cases had the lowest preference. This priority ranking did not change when we set (w1 = w2 = w3 = w4) > (w5 = w6), and when w1 = w2 = w3 = w4 = w5 = w6 also, assuming that the different criteria have the same weight for the purposes of the overall assessment of each area. The priority order of the park areas, on the other hand, varied substantially when the highest importance was attributed to criterion C5. In this last case, the area that presented the greatest preferability was D. This was also confirmed when we assumed (w5 = w6) > (w1 = w2 = w3 = w4), i.e., when greater importance was assigned to the criteria that expressed the "extrinsic" quality (use value) of the park areas.

The overall priority of the homogeneous zones identified in the park area is, therefore, the following:

- Zone A: 4;
- Zone B: 3;
- Zone C: 1;
- Zone D: 2.

The above classification represents the least "sensitive" ranking to variations in the weighting system of the evaluation criteria. In it, zone A, "natural environment", was the one with the highest complex social value, in which it is appropriate to preserve or increase the naturalistic values, excluding any type of transformation. The complex social value of zone B, "semi-natural environmen", was lower, for which a generalized protection must consequently be envisaged, which can result in the inhibition of activities that involve irreversible ecosystems modifications. In zones D, "urban environment", and C, "agro-forestry and agrarian environment", with a complex social value, medium and low transformations, respectively, can be introduced, whose impacts are not incompatible with the qualitative components of the area (natural, historical and cultural resources).

4.4. Activities Planned for the Natural Park Establishment

The knowledge of the protection degrees assigned to the homogeneous zones of the park area allows us to establish the types of activities that can be carried out in them.

These activities are indicated in Table 7 in the form of active and passive requirements for the individual zones.

Table 7. Activities that can be implemented for the establishment of the park and positive or negative prescriptions for the individual homogeneous areas (+ permitted activity, − prohibited activity).

Activities	Zone			
	A	B	C	D
Construction of buildings and products in general	−	−	+	+
Construction of new infrastructures	−	−	+	+
Installation of visual technological systems	−	−	+	+
Construction of systems for collective and/or tourist use	−	+	+	+
Use of synthetic fertilizers	−	−	+	+
Withdrawals from surface or underground water bodies	−	−	+	+
Traffic and vehicle parking	−	+	+	+
Collection of minerals for production purposes	−	+	+	+
Modification of water regulation, riverbeds, or reservoirs	−	−	+	+
Extension and adaptation of driveways	−	+	+	+
Crop transformations	−	−	+	+
Reforestation	−	+	+	+
Exploitation of hydrothermal and mineral springs	−	+	+	+
Agronomic practices	−	+	+	+
Farming	−	+	+	+
Industrial activities	−	−	+	+

4.5. Design Alternatives

The design alternatives for the study area were defined by taking into account indications contained in the Territorial Urban Plan of the Mountain Community "Valle dell'Irno", and proposals made by local public administrations and authorities, as well as by naturalistic associations.

The types of intervention hypothesized are compatible with the activity categories that can be implemented in the homogeneous areas.

Three design alternatives were considered.

Articulated into types of intervention that compose them, the design alternatives are indicated in Table 8.

Alternative 1 is the one that most closely matches the status quo. The planned interventions essentially refer to environmental protection (environmental control and restoration) and resource use. The interventions related to resource use are aimed mainly to satisfy a demand with naturalistic reasons.

Alternative 2 provides, in close connection with the protection measures necessary to safeguard the natural environments present in the park area, a set of resource enhancement interventions, aimed at encouraging the correct use of resources by visitors because they exercise demand segments with multiple motivations (naturalistic, cultural, sporting, recreational, etc.).

Alternative 3 differs from the previous ones because it aims to increase, to a greater measure than the others, the production capacity of the park area, enhancing its potential through support structures for existing economic activities and through the promotion of new productive activities.

Table 8. Summary of design alternatives: green cells represent the possibility of admitting the intervention in the area.

Green-filled cells are marked with **X**; empty cells are blank.

Type of Intervention	Alternative 1 (Zones)				Alternative 2 (Zones)				Alternative 3 (Zones)			
	A	B	C	D	A	B	C	D	A	B	C	D
Pollution control and monitoring	X	X			X	X			X	X		
Quarry restoration			X	X			X	X			X	X
Landfill remediation		X	X			X	X			X		
Faunistic farms											X	
Animal shelters					X	X				X		
Equipped green areas							X	X			X	X
Horse riding routes						X						
Nature trails		X	X		X	X	X			X		
Naturalistic observers		X				X				X		
Food points							X				X	
Hilly lakes						X						
Support structures for zootechnics						X					X	
Areas equipped for grazing						X						
Harvesting center for agricultural products											X	
Chestnuts processing and marketing plant												
Social center for the elderly								X				
Visitor and information center				X			X				X	X
Museum of the territory							X					
Laboratory for scientific and didactic activities						X						
Parking lots							X				X	
Signage	X	X			X							
Typical products promotion center						X						
Forest trails						X						
Camping						X						
Hotel											X	
Pension											X	
Horse riding facilities						X						
Oil mill												X

4.6. Order of Preferability for the Design Alternatives: Optimal Alternative Choice

A priority ranking among design alternatives must be constructed by first evaluating each alternative with respect to defined criteria and objectives. The alternative ordering is then determined with the application of specific multicriteria analysis, after assigning weights to the criteria/objectives considered.

The design alternatives defined for the park area under study were evaluated using the following criteria:

- Environmental protection criterion. This searches for the solution that minimizes the loss or compromise of resources and irreproducible emergencies. To make this criterion operational, an indicator of the consistency of environmental resources that will be destroyed or compromised with the project realization must be used.
- Ethical/social criterion. This is the result of searching for the solution that produces the highest increase in occupation degree in the gravitation area of the park. The increase in the number of employees produced by the activities in the park area with the project realization should be taken as an indicator of this.
- Criterion of economic valorization territory. This can be expressed through the solution searching which corresponds to the highest economic benefit for the community. The internal economic rate of return of the project can be used as an indicator of this.
- In other words, for the park area considered, the identification of the optimal alternative must be achieved with the search of the design solution capable of minimizing the environmental cost and, at the same time, maximizing the social and economic benefits.

In the application of the above criteria, both the environmental cost and the social and economic benefit were envisaged for each alternative outlined.

The evaluation result is shown in Table 9. The latter table is an ordinal impact matrix with indexes ranging from 1 to 3 (1 is the minimum preferability; 3 is the maximum preferability).

Table 9. Order of preferability for the design alternatives.

Alternatives	Minimum Environmental Costs	Maximum Social Benefits	Maximum Economic Benefits
1	3	1	1
2	2	2	2
3	1	3	3

Additionally, for the definition of the overall priority of the three design alternatives, the regime analysis developed by Nijkamp and Hinloopen was applied, already used in the analysis to define the ordering of the homogeneous areas identified in the park area.

Table 10 shows the preferability rankings of the alternatives according to the possible ordinal systems of weighting of the criteria.

Table 10. Preferability rankings of the design alternatives taking into account the possible ordinal systems of weighting of the criteria.

Weights			Alternatives		
w1	w2	w3	1	2	3
2	2	2	1	2	3
3	2	1	2	1	3
3	1	2	2	1	3
2	3	2	1	2	3
1	3	2	1	2	3
2	3	1	2	1	3
2	2	3	1	2	3
1	2	3	1	2	3
2	1	3	1	2	3
3	3	2	1	2	3
2	3	3	1	2	3
3	2	3	1	2	3

Alternative 3 expressed the highest preferability, and therefore, the highest complex social value, in correspondence with every possible combination of weights assigned to the three project objectives. In general, Alternative 2 was less preferable, followed by Alternative 1. The latter became preferable to Alternative 2 when the greatest importance was assigned to the objective of reducing the environmental cost, i.e., when the objective of greater importance was the maximization of social benefits. In all other cases, Alternative 1 was less preferable than Alternative 2. This was also confirmed when we set w1 = w2 = w3.

The ordering representing the overall priority of the project alternatives hypothesized for the park area is, therefore: Alternative 1 equal to overall priority 1; Alternative 2 equal to overall priority 2; Alternative 3 equal to overall priority 3.

The highest overall priority was expressed by Alternative 3, which consequently constitutes the design solution capable of optimally integrating environmental protection objectives with the objectives of social and economic enhancement of the territory. Alternatives 2 and 1 are less preferable, the reciprocal position of which in the system may nevertheless undergo changes in the event that environmental and ethical/social considerations require less importance to be attributed to the objective of economic development of the territory.

5. Concluding Remarks

The management of sustainable development is an open and current issue.

In the design of natural parks, the examination and definition of the compatibility between qualitative and quantitative growth profiles of the park system, as well as the intervention choice to be implemented to achieve the protection level and enhancement of resources, involve a necessary expansion of the evaluation framework and parameters.

A balanced and sustainable development strategy of the park area implies, in fact, the search for solutions capable of satisfying diversified needs, attributable to economic needs and to aspects more directly related to ecological and social quality. Consequently, the use of evaluation methodologies that lead above all to the identification of the different components of the value of the resources and then to the analysis of the interdependencies between components is indispensable in order to group them according to a multidimensional scheme. A scheme that adequately reconnects, in order to arrive at an overall result, economic, social and environmental valuations.

The complex social value collects the variables of different nature that form the value of the resources and emergencies present in the area delimited for the constitution of the park. It reflects the weight of the economic, ethical–social and ecological variables of the resources being evaluated. Therefore, it understands and expresses the multiple expectations of the community with regard to the use of natural and historical–cultural resources within the planning and implementation processes of the reorganization and growth projects of the territory.

However, the application of the complex social value cannot ignore the multicriteria analysis—the evaluation techniques belonging to this family—taking into account the plurality of quantitative and qualitative components that make up the value of environmental resources. These techniques also make it possible to recognize and explain at qualitative–quantitative scales the complex of direct and indirect impacts that can be generated by possible interventions in the economic, social and ecological contexts of reference. Unlike traditional evaluation techniques, they appear to be capable of translating, in a global judgment, the multidimensionality of the aspects and the interdependence of the variables on which the choice of the solutions to be implemented depends in optimal terms.

The prospects for the use of "complex" assessments—undoubtedly linked to the development of landscape planning but even more, in general, to the implementation of planning and requalification policies of the territory—are nevertheless still conditioned by the inadequacy of methodological tools. This concerns, in particular, the processing of data and the objectification of information of a qualitative nature, as well as, above all, the problem of assessing "environmental quality".

Some limitations of the approach used in this study are represented by the regulations in force in natural parks and protected areas. Each individual country can provide a legislation that has the defect of sometimes being confused with respect to the European Community legislation or directives from worldwide coordination authorities, although in recent years, there has been numerous legislative novelties on the subject. National regulations can in fact be based on a purely static–conservative conception of the wooded and natural heritage, rather than based on operations to safeguard and transform the soil, as well as to protect the environment. Furthermore, the regulation of protection and enhancement activities, within the individual areas to be protected, may or not enable the establishment of specific management authorities, which, among other things, have the task of harmonizing protection plans and actions with territorial planning guidelines. These are all issues that, very often, determine conflicting aspects between environmental protection and economic operators involved in the processes of territorial transformation and redevelopment.

Author Contributions: V.D.G. contributed to the conceptualization and supervision; P.D.P. contributed to the formal analysis and methodology; P.M. and F.T. contributed to the investigation and supervision; F.P.D.G. contributed to the data curation, software and validation results. All authors have read and agreed to the published version of the manuscript.

Funding: This research received no external funding.

Conflicts of Interest: The authors declare no conflict of interest.

References

1. Forte, F.; Girard, L.F.; Nijkamp, P. Smart Policy, Creative Strategy and Urban Development. *Stud. Reg. Sci.* **2006**, *35*, 947–963. [CrossRef]
2. Forte, F.; Girard, L.F. Creativity and new architectural assets: The complex value of beaty. *Int. J. Sustain. Dev.* **2009**, *12*, 160–191. [CrossRef]
3. Del Giudice, V.; De Paola, P.; Forte, F. Valuation of historical, cultural and environmental resources, between traditional approaches and future perspectives. In *Green Energy and Technology*; Springer: Berlin/Heidelberg, Germany, 2018; pp. 177–186.
4. Forte, F.; Del Giudice, V.; De Paola, P.; Del Giudice, F.P. Cultural heritage and seismic disasters: Assessment methods and damage types. In *Green Energy and Technology*; Springer: Berlin/Heidelberg, Germany, 2021; pp. 163–175.
5. Manganelli, B.; Vona, M.; De Paola, P. Evaluating the cost and benefits of earthquake protection of buildings. *J. Eur. Real Estate Res.* **2018**, *11*, 263–278. [CrossRef]
6. Campbell, H. Planning: An idea of value. *Town Plan. Rev.* **2002**, *73*, 271–288. [CrossRef]
7. Girard, L.F.; Cerreta, M.; De Toro, P.; Forte, F. The human sustainable city: Values, approaches and evaluative tools. In *Sustainable Urban Development. The Environmental and Assessment Methods*; Deakin, M., Mitchell, G., Nijkamp, P., Vreeker, R., Eds.; Routledge: London, UK, 2007; Volume 2, pp. 65–93.
8. Girard, L.F. *Risorse Architettoniche e Culturali: Valutazioni e Strategie di Conservazione*; FrancoAngeli: Milano, Italy, 1987.
9. Girard, L.F.; Nijkamp, P. *Energia, Bellezza e Partecipazione: La Sfida della Sostenibilità. Valutazioni Integrate tra Conservazione e Sviluppo*; FrancoAngeli: Milano, Italy, 2004.
10. Zeleny, M. Multiple criteria decision-making: Eight concepts of optimality. *Hum. Syst. Manag.* **1998**, *17*, 97–107.
11. Zeleny, M. *Human Systems Management: Integrating Knowledge, Management and Systrems*; World Scientific Publishers: Hackensack, NJ, USA, 2005.
12. Giampietro, M.; Allen, T.F.H.; Mayumi, K. Science for governance: The implications of the complexity revolution. In *Interfaces between Science and Society*; Gutmaraes-Pereira, A., Guedes-Yaz, S., Tognetti, S., Eds.; Greenleaf Publishing: Sheffield, UK, 2006; pp. 82–99.
13. Munda, G. *Social Multi-Crireria Evaluarion for a Sustainable Economy*; Springer: Berlin/Heidelberg, Germany, 2008.
14. Funtowicz, S.O.; Martinez-Alier, J.; Munda, G.; Ravetz, J. Multi-criteria-based environmental policy. In *Implementing Sustainable Development*; Abaza, H., Baranzini, A., Eds.; UNEP/Edward Elgar: Cheltenham, UK, 2002; pp. 53–77.
15. Munda, G. Social multi-criteria evaluation: Methodological foundations and operational consequences. *Eur. J. Oper. Res.* **2004**, *158*, 662–677. [CrossRef]
16. Larner, W.; Le Heron, R. The spaces and subjects of a globalising economy: A situated exploration of method. *Environ. Plan. D Soc. Space* **2002**, *20*, 753–774. [CrossRef]
17. Mingers, J.; Rosenhead, J. *Rational Analysis for a Problematic World Revisited: Problem Structuring Merhods for Complexity, Uncertainty and Conflict*; Wiley: Chichester, UK, 2001.
18. Cats-Baril, W.L.; Huber, G.P. Decision support systems for Ill-structured problems: An empirical study. *Decis. Sci.* **1987**, *18*, 350–372. [CrossRef]
19. Rittel, H.; Webber, M. Dilemmas in a general theory of planning. *Policy Sci.* **1973**, *4*, 155–169. [CrossRef]
20. Rosenhead, J. Controversy on the streets: Stakeholder workshops on a choice a carnival route. In *Planning under Pressure: The Strategic Choice Approach*; Friend, J., Hickling, A., Eds.; Butterworth-Heinemann: Oxford, UK, 2005; pp. 298–302.
21. Schon, D.; Rein, M. *Frame Reflection: Toward the Resolution of Intractable Controversies*; Basic Books: New York, NY, USA, 1994.
22. Strauss, K. Re-Engaging with rationality in economic geography: Behavioural approaches and the importance of context in decision-making. *J. Econ. Geogr.* **2008**, *8*, 137–156. [CrossRef]
23. Alexander, E.R. (Ed.) *Evaluation in Planning. Evolution and Prospects*; Ashgate: Aldershot, UK, 2006.
24. Lichfield, N. *Community Impact Evaluation*; UCL Press: London, UK, 1996.
25. Deakin, M.; Mitchell, G.; Nijkamp, P.; Vreeker, R. (Eds.) *Sustainable Urban Development. The Environmental Assessment Merhods*; Routledge: London, UK, 2007; Volume 2.
26. Miller, D.; Patassini, D. (Eds.) *Beyond Benefit Cost Analysis. Accounting for Non-Market Values in Planning Evaluation*; Ashgate: Aldershot, UK, 2005.
27. Sherrouse, B.C.; Semmens, D.J.; Clement, J.M. An application of Social Values for Ecosystem Services (SolVES) to three national forests in Colorado and Wyoming. *Ecol. Indic.* **2014**, *36*, 68–79. [CrossRef]
28. Fulgencio, H. Social value of an innovation ecosystem: The case of Leiden Bioscience Park, The Netherlands. *Int. J. Innov. Sci.* **2017**, *9*, 355–373. [CrossRef]

29. Fagerholm, N.; Raymond, C.M.; Olafsson, A.S.; Brown, G.; Rinne, T.; Hasanzadeh, K.; Broberg, A.; Kyttä, M. A methodological framework for analysis of participatory mapping data in research, planning, and management. *Int. J. Geogr. Inf. Sci.* **2021**, 1–28. [CrossRef]

30. Ribera, F.; Nesticò, A.; Cucco, P.; Maselli, G. A multicriteria approach to identify the highest and best use for historical buildings. *J. Cult. Herit.* **2020**, *41*, 166–177. [CrossRef]

31. Guarini, M.R.; Morano, P.; Sica, F. Historical school buildings. A multicriteria approach for urban sustainable projects. *Sustainability* **2020**, *12*, 1076. [CrossRef]

32. Guarini, M.R.; D'Addabbo, N.; Morano, P.; Tajani, F. Multi-criteria analysis in cmpound decision processes: The AHP and the architectural competition for the Chamber of Deputies in Rome (Italy). *Buildings* **2017**, *7*, 38. [CrossRef]

33. Morano, P.; Locurcio, M.; Tajani, F.; Guarini, M.R. Fuzzy logic and coherence control in multi-criteria evaluation of urban redevelopment projects. *Int. J. Bus. Intell. Data Min.* **2015**, *10*, 73–93. [CrossRef]

34. Saaty, T.L.; De Paola, P. Rethinking Design and Urban Planning for the Cities of the Future. *Buildings* **2017**, *7*, 76. [CrossRef]

35. Giaoutzi, M.; Nijkamp, P. *Decision Support Models for Regional Sustainable Development*; Avebury: London, UK, 1994.

36. Nijkamp, P.; Hinloopen, J. A sensitivity analysis of multicriteria choice methods. *Energy Econ.* **1989**, *11*, 293–300.

37. Nijkamp, P.; Hinloopen, J. Qualitative multiple criteria choice analysis. *Qual. Quant.* **1990**, *24*, 37–56.

38. Girard, L.F.; Nijkamp, P. *Le Valutazioni per lo Sviluppo Sostenibile Della Città e del Territorio*; Franco Angeli: Milano, Italy, 1997.

39. Nijkamp, P.; Rietveld, P.; Voogd, H. *Multicriteria Evaluation in Phisical Planning*; North Holland: Amsterdam, The Netherlands, 1990.

40. Del Giudice, V.; De Paola, P. Undivided real estate shares: Appraisal and interactions with capital markets. *Appl. Mech. Mater.* **2014**, *584–586*, 2522–2527. [CrossRef]

41. Del Giudice, V.; De Paola, P. The effects of noise pollution produced by road traffic of Naples Beltway on residential real estate values. *Appl. Mech. Mater.* **2014**, *587–589*, 2176–2182. [CrossRef]

42. Del Giudice, V.; De Paola, P.; Del Giudice, F.P. COVID-19 Infects Real Estate Markets: Short and Mid-Run Effects on Housing Prices in Campania Region (Italy). *Soc. Sci.* **2020**, *9*, 114. [CrossRef]

43. Del Giudice, V.; Massimo, D.E.; De Paola, P.; Del Giudice, F.P.; Musolino, M. Green buildings for post carbon city: Determining market premium using spline smoothing semiparametric method. In *Smart Innovation, System and Technologies*; 178 SIST; Springer Nature: Berlin, Germany, 2021; pp. 1227–1236.

44. De Paola, P.; Del Giudice, V.; Massimo, D.E.; Del Giudice, F.P.; Musolino, M.; Malerba, A. Green buildings market premium: Detection through spatial analysis of real estate values. A case study. In *Smart Innovation, System and Technologies*; 178 SIST; Springer Nature: Berlin, Germany, 2021; pp. 1413–1422.

45. Del Giudice, V.; De Paola, P.; Bevilacqua, P.; Pino, A.; Del Giudice, F.P. Abandoned Industrial Areas with Critical Environmental Pollution: Evaluation Model and Stigma Effect. *Sustainability* **2020**, *12*, 5267. [CrossRef]

*applied sciences*

MDPI

Review

Sustainable Cross-Laminated Timber Structures in a Seismic Area: Overview and Future Trends

Antonio Sandoli *, Claudio D'Ambra, Carla Ceraldi, Bruno Calderoni and Andrea Prota

Department of Structures for Engineering and Architecture, University of Naples Federico II, via Claudio 21, 80125 Naples, Italy; claudio.dambra@uninat.it (C.D.); ceraldi@uninat.it (C.C.); calderon@uninat.it (B.C.); aprota@uninat.it (A.P.)
* Correspondence: Antonio.sandoli@unina.it

Abstract: Cross-laminated timber (CLT) buildings are recognized as a robust alternative to heavy-weight constructions, because beneficial for seismic resistance and environmental sustainability, more than other construction materials. The lightness of material and the satisfactory dissipative response of the mechanical connections provide an excellent seismic response to multi-story CLT buildings, in spite of permanent damage to timber panels in the connection zones. Basically, CLT constructions are highly sustainable structures from extraction of raw material, to manufacturing process, up to usage, disposal and recycling. With respect to other constructions, the potential of CLT buildings is that their environmental sustainability in the phases of disposal and/or recycling can be further enhanced if the seismic damage in structural timber components is reduced or nullified. This paper reports a state-of-the art overview on seismic performance and sustainability aspects of CLT buildings in seismic prone regions. Technological issues and modelling approaches for traditional CLT buildings currently proposed in literature are discussed, focusing the attention on some research advancements and future trends devoted to enhance seismic performance and environmental sustainability of CLT buildings in seismic prone regions.

Keywords: CLT buildings; sustainability; environment; seismic behavior; traditional connections; low-damage connections

Citation: Sandoli, A.; D'Ambra, C.; Ceraldi, C.; Calderoni, B.; Prota, A. Sustainable Cross-Laminated Timber Structures in a Seismic Area: Overview and Future Trends. *Appl. Sci.* **2021**, *11*, 2078. https://doi.org/10.3390/app11052078

Academic Editors: Elmira Jamei and Panagiotis G. Asteris

Received: 26 January 2021
Accepted: 23 February 2021
Published: 26 February 2021

Publisher's Note: MDPI stays neutral with regard to jurisdictional claims in published maps and institutional affiliations.

1. Introduction

Timber-based structures are gaining popularity in residential and non-residential constructions marked in the last two decades, also in those countries not prone to the use of wood. This development is mainly due to the introduction of the high efficiency engineered and sustainable timber products such as Glue-Lam, cross-laminated timber (CLT) and Laminated Veneer Lumber (LVL) used for one and two-dimensional elements.

Cross-lam panel, originated in Austria in the early of 1990s, represents one of the most diffused mass timber products in Europe for low and mid-rise timber buildings. It has been estimated that the global annual production of CLT is growing exponentially in Europe, it was 25,000 m^3 in 1996, 340,000 m^3 in 2010, 650,000 m^3 in 2015, up to 1.2 million of m^3 in 2020 [1,2] (Figure 1). Nevertheless, an annual increase of the manufactured CLT has been also recorded in the United States, Canada, Australia, Japan and New Zealand [3–5].

Together with its excellent seismic resistance, environmentally-friendly and eco-sustainability properties have made wood a particularly appreciated building material. By contrast with other mass timber structures, boards used for CLT panels can be obtained from small-diameter trees: such trees have low commercial value and can be beneficial in maintaining a healthier forests habitat, because less vulnerable to wildfires [6].

Furthermore, aspects related to life-cycle assessment encouraged the use of CLT in building market: CLT, as well as other timber-based products, has better energy saving and carbon reduction performance than other traditional building materials. Boriesson et al. [7] demonstrated that compared with reinforced concrete-framed buildings, wooden-framed constructions consume approximately 80% more energy during the material production stages and release about 100–200% more. With regards to CLT, Guo et al. [8] pointed out that, on a national level, a seven-story CLT building may result in a 29.4% energy saving (which is equivalent to 24.6% carbon reduction) when compared with reinforced concrete building in the operational stage. Liu et al. [9] compared the consumption between two seven-story buildings constructed using reinforced concrete and CLT panels: they demonstrated as the energy consumption for heating and cooling was 338 MJ/m^2 and 231 MJ/m^2 per annum for reinforced concrete and CLT, respectively.

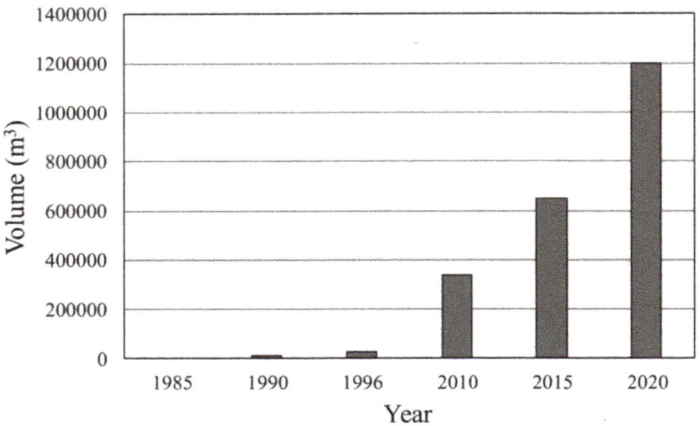

Figure 1. Volume in m^3 of cross-laminated timber (CLT) manufacture in Europe.

The great potential for CO_2 reduction achievable by wood structure buildings has been also proved. The CO_2 emissions of houses with a concrete structure are 850 kg/m^2, compared with 250 kg/m^2 for wooden structures and 450 kg/m^2 for light-weight steel structures [10].

As far as the seismic resistance goes, the high strength-to-weight ratio allow medium and high-rise CLT buildings capable of withstanding high-intensity earthquakes [2,11], despite the lack of code prescriptions to design seismic-resistant CLT structures at national or international levels. The majority of prescriptions for the in-plane and out-of-plane checks of the panels at ultimate and serviceability limit states, as well as seismic design rules for multi-story buildings, can be found in literature papers or in manuals only. Nevertheless, the need to provide practical rules for design practice kicked off the update of some standards in Europe (Eurocode 5 [12] and Eurocode 8 [13]), even if these are not yet concluded [14,15].

Experimental, theoretical and numerical studies highlighted a satisfactory seismic performance of CLT buildings. A shake table test performed on full-scale multi-story buildings proved the capability of surviving high-intensity earthquakes, up to 0.82 g of peak-ground acceleration for a seven-story building [16,17] and up to 1.52 g for a two-story building [18], depending on lightness of material and behavior of mechanical connections.

Local tests on connections showed a significant energy dissipation capacity, if (correctly) designed to prevent brittle failure modes [19–23]. Mechanical connections, placed in the panel to panel and panel to foundation contact zones, named as hold-down (HD) and angle brackets (AB), consist of thin steel plates nailed or screwed to timber panels and bolted to foundation: the first resist to tension forces and are used against uplift, while the second resist to shear forces and are used to withstand sliding. HD and AB play a crucial role on the seismic performance of CLT buildings because they are the main source of seismic energy dissipation.

In the perspective of performance-based seismic design (PBD), HDs and ABs may be regarded as "traditional" systems although introduced by a few years. Their failure mechanisms involve high damage to timber structural components after an earthquake, enough to be replaced by alternative systems. High damaging is in conflict with PDB philosophy, which requires high seismic performance and low damaging contemporarily satisfied by the structures.

Serious damage to timber parts affects also the sustainability related to the final part of life cycle of the construction, and in particular phases of the disposal and/or recycling of material. Basically, thanks to eco-sustainability of material and attention of manufacturers to produce environmentally certified products, CLT buildings are highly-sustainable constructions. However, the sustainability can be furtherly improved if design approaches oriented to seismic damage reduction on timber components are developed.

To this purpose, the first studies aimed at developing an "integrated design" between seismic-resistant and more sustainable timber constructions are under development; the main scope is that of introducing low-damage connections capable of (i) concentrating the seismic dissipation in replaceable connection systems and of (ii) reducing seismic damages on timber components [24–26].

This paper presents a state-of-the-art on the seismic behavior of traditional CLT buildings, focusing on the main research advancements aimed at enhancing both seismic performance and environmental sustainability. The first sections of the article are dedicated to the description of the technological aspects and seismic behavior of CLT buildings. In the following sections the different modelling strategies for multi-story buildings available in the scientific literature are illustrated and compared critically. The state-of-the-art allowed pros and cons of CLT constructions with regards to sustainability to be highlighted and to discuss critically (in the finale of the paper) about the new and future research trends concerning the seismic performance and environmental sustainability enhancing of CLT buildings.

2. Cross-Laminated Timber (CLT) Panels

CLT panel is a two-dimensional mass timber element composed of an appropriate number of layer of boards. The layers are made with adjacent boards or lamellae (usually not glued on the edges), rotated around at a right angle one each other and glued among them (Figure 2).

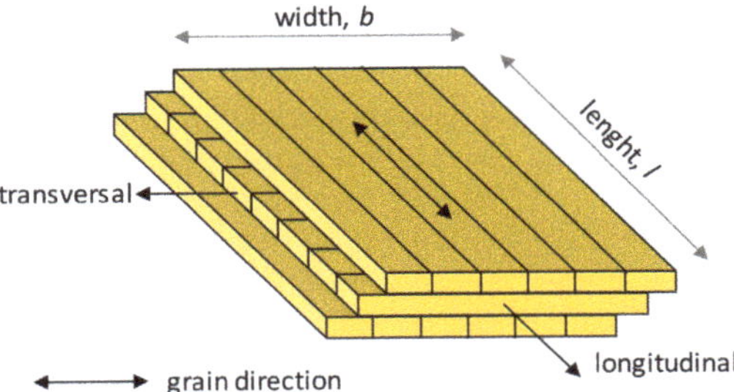

Figure 2. CLT panel lay-out.

Panel dimensions are variable as regards thickness (t), width (b) and length (l). The length can in theory be unlimited because the individual boards are finger-joined to create long boards. The number of layers is odd and vary from a minimum of three to a maximum of nine. In Table 1 are indicated the common dimensions used in building practice.

Table 1. Range of the geometrical characteristics of cross-laminated timber (CLT) panels.

Parameter	Commonplace
Thickness (t)	60–300 mm
Width (b)	1.20–4.80
Length (l)	up to 30 m
No. of layers	3–5–7–9

The raw material consists of timber boards that have been strength graded according to standards (e.g., EN-14081-1 in Europe [27]). The cross-section of CLT usually comprises boards of the same strength in the two main direction, here indicated as "longitudinal" (axis of major strength and stiffness) and "transversal" (axis of minor strength and stiffness) in Figure 2. To make the most of the panel strength, wood with higher strength can be used in the main direction of the load.

As for the majority of engineered wood products, also for CLT panels soft-wood species are mainly used, such as spruce or less frequently larch or douglasia (that ensure greater durability to the panel). Depending on geographic area, the boards arrangement can be different [11], as represented in Figure 3.

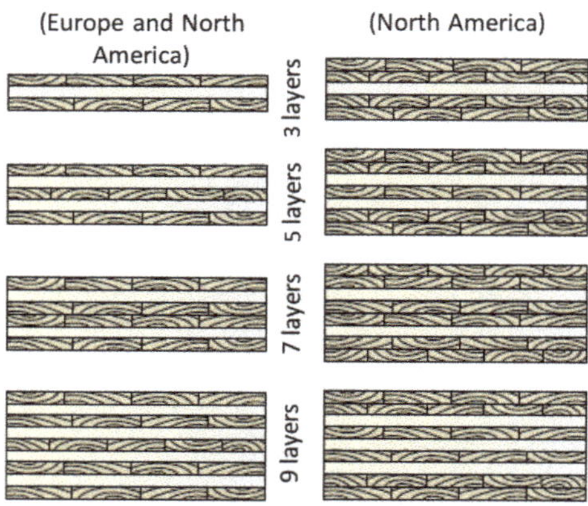

Figure 3. Number of layers and boards arrangement in CLT panels (image from [11]).

The cross-layered morphology provides high structural properties to CLT members, that can compete with other more traditional structural materials adopted for medium and high-rise buildings. The main advantages of CLT panels regard the structural features, thermal and sustainable aspects, as summarized in the following:

- In relation to their own-weight, CLT panels have a higher load-bearing capacity than most other construction materials: this means that high-rise buildings can be constructed with reduced masses.
- The cross-layered nature provides in-plane and out-of-plane load-carrying capacity and can be used either for vertical shear-walls (membranes) or floors (plates), as well as for other structural applications (bridges, sport arenas, curved elements, etc.).
- High in-plane strength and stiffness, even if differentiated between the longitudinal and transversal directions of the panel.
- Good dimensional stability against moisture variations: crossed layers reduce the shrinkage and swelling phenomena with respect to other timber-based products.
- Excellent energy efficiency and capacity of storing moisture and thermal energy.
- Ecofriendly, recyclable and renewable construction material with long service life (when correctly protected against moisture).
- Workability, slenderness and different sizes do not limit architectural design.

Despite the significant use in construction practice, CLT panels are not yet codified in national or international standards. In Europe, the European Technical Assessments (ETAs) represents the main reference document that codifies the CLT products indicating their specific properties for structural applications [28]. In compliance with ETAs, CLT panels are made with boards having a strength class corresponding to C16, C24 and C30.

Instead, the National Annexes to the European standards [29,30] specify the design rules and checks, although they have not been yet included in the Eurocode 5 [12]. In Europe, the revised Italian Technical Document CNR DT 2016/2018-R1 [31] contemplates the use of the CLT and indicates its main characteristics for structural purposes.

In the United States, specific properties of CLT products and their structural uses for buildings are included in the International Building Code [32] while for strength and fire resistance checks it refers to the National Design Specification (NDS) for Wood Construction [33].

In Japan, the Building Standard Law has included the possibility of CLT constructions since 2016, but without providing design and checks rules. Instead, the reference guidelines for CLT constructions are contained in a specific manual released by the Japan CLT Association [34]. Also Australia and New Zealand lack specific design codes, but can rely only on guideline prescriptions [35].

In the meantime, engineers and researchers can trust on not-codified rules included in books or in manuals released by research groups or manufacturers, which offer recommendations based on experimental and numerical results. Useful documents are the manuals released by FP-Innovation in Canada [11], the Swedish one [2] or that issued by proHolz-Austria in 2015 [36], all contain indications for the structural design and strength checks at ultimate and service limit states for CLT panels.

3. Traditional CLT Buildings

3.1. Technology

Low and mid-rise traditional CLT buildings are made with CLT panels used for both vertical shear-walls and horizontal floors. The panels are prefabricated off-site and connected on the building site using metal connectors such as hold−downs, angle brackets, screws and nails.

Depending on their arrangement, the shear-walls can be divided in two groups: (i) single (or monolithic) and (ii) segmented (or coupled) walls. In the case (i) the wall has high width-to-height ratio, with width equal to plant dimension of the building and length corresponding to its inter-story height. In the case (ii) the wall is composed by narrow panels joined to each other by means of vertical step joints [21,37], and width-to-height ratios greater than 0.20 are advisable [31].

CLT traditional buildings are constructed using the "platform" technique, i.e., the vertical continuity of the walls is interrupted at each story by horizontal panels (floors) included, as a "sandwich", between two consecutive vertical panels (Figure 4). At the ground floors the vertical panels are directly placed on a reinforced concrete foundation or a horizontal timber beam interposed between foundation and vertical panels. Low-rise buildings can be also built-up adopting the "balloon" technique (Figure 5), where the vertical continuity of the walls is not interrupted by the floors at each story, but they are connected on the internal side of the vertical walls by means of specific connection devices [38].

Panel to panel and panel to foundation connections consist of mechanical steel to timber connections, made with punched cold-formed thin steel plates-typically hold-downs (HD) and angle brackets (AB)—fastened to the timber panels with ring shanks nails or screws and bolted to the foundations (Figure 4).

Being characterized by a high level of prefabrication, several connection-types are needed to connect the panels among them (in addition to HDs and ABs), such as metal fasteners between the orthogonal walls, metal fasteners to join the floor panels or to connect segmented walls against sliding (Figure 4). Connections between floor panels and between segmented walls can be realized with different types of fasteners arranged differently, typically overlapped joints with inclined screws running at 45°, half-lap joints with vertical screws or other solutions which can be found in [2,11].

Door and window openings are obtained as a function of the type of the vertical walls arrangement: by cutting the zone interested by the opening in the case of monolithic walls, or by assembling multiple elements in the case of segmented walls. In the first case the lintels, i.e., horizontal CLT panels above the openings, have the same cross layout of the vertical panels, while in the second case the layout can be also different.

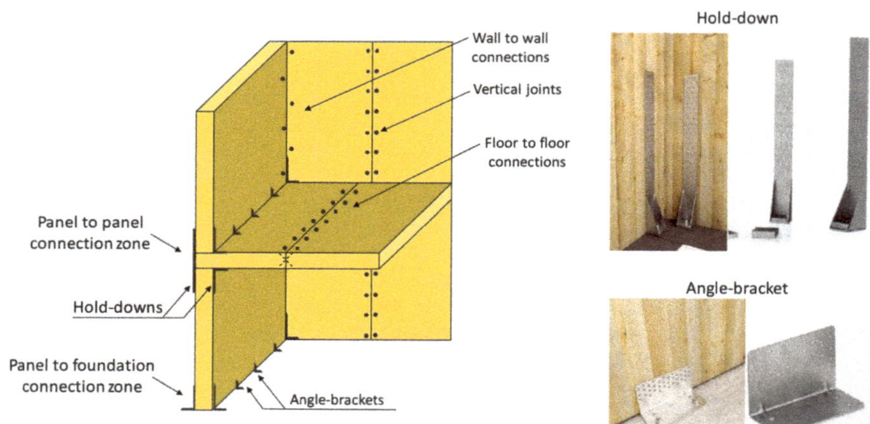

Figure 4. Typical connections for CLT buildings constructed with platform technique.

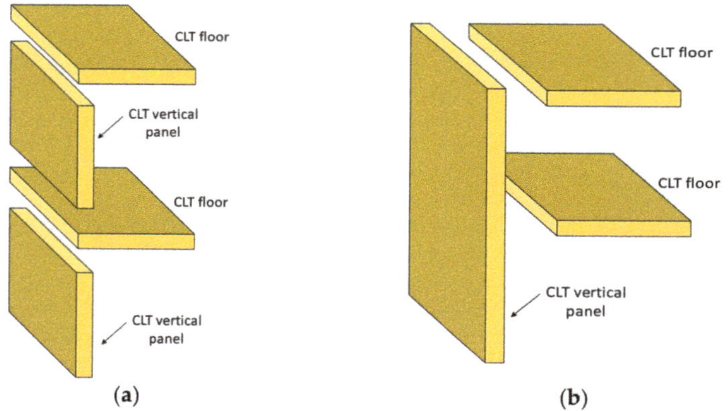

Figure 5. Constructional techniques for CLT buildings: (**a**) platform, (**b**) balloon.

3.2. Seismic Behavior

Despite the high number of connections, CLT buildings exhibit a favorable box-like behavior under combined seismic actions and gravity loads.

With regards the gravity loads patterns only, particular precautions on connection systems are not necessary. Due to the platform constructive technique, the compressive forces at the interface between the vertical panels and the interposed floor panels (in the sequel referred as "panel to panel" contact) are transferred by contact. Such compressive forces load in orthogonal to grain direction the floor panels in which a low-strength of material is expected, because the orthogonal to grain compressive strength ($f_{c,90}$) is about $1/10$ of the parallel to grain compression strength ($f_{c,0}$). This issue sets limits also with respect to the maximum height of the buildings, which should be at most 10 storey in height in order to avoid high stress levels on the floor panels. On the contrary, the panels cross-section dimensions typically used for low and mid-rise buildings are overestimated with respect to the amount of the gravity loads. It is worth underlining that, in case of balloon constructive technique the issue of orthogonal to grain loading of panel floors does not arise because they do not interrupt the vertical continuity of the wall.

Seismic actions, instead, entail a more complex load path which involves the mechanical connections in the panel to panel and the panel to foundation contact zones, as schematically represented in Figure 6. Hold-downs are devoted to face uplift and then withstand principally the tension forces due to the overturning moment, while angle-brackets contrast principally the sliding and then absorb the shear forces. While the timber-to-timber contact ensures the transferring of compression forces due to horizontal actions at the panel-to-panel or panel-to-foundation interfaces.

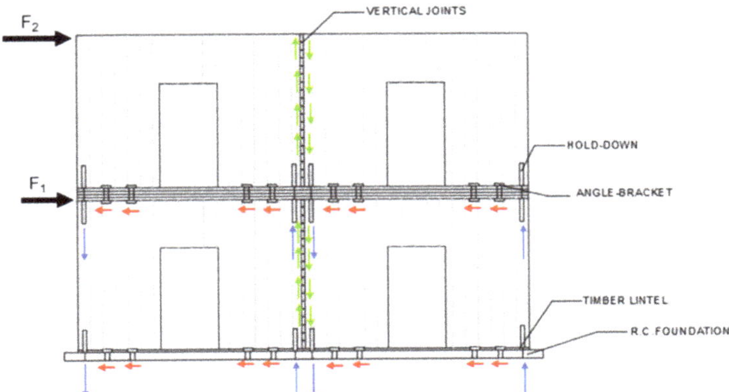

Figure 6. Seismic action paths for a CLT shear-wall.

A full box-like behavior of the construction is achieved if CLT floors are sufficiently rigid in own-plane. This assumption could be compromised (and then requires verifications) especially in the case of a floor realized by adjacent panels connected by screws. The relative elastic sliding which occurs among the panels in the joined zones could make the floors not adequately rigid in the horizontal plane; this produces a transferring of seismic actions among the seismic-resistant walls as a function of their tributary area, encouraging potential out-of-plane behavior of the vertical panels themselves. In any case, this phenomenon can be prevented by stiffening the floor by means of reinforced concrete slabs connected with steel fasteners to the timber panels. On the contrary, in the case of high-intensity earthquakes, the dissipation in the panel-to-panel connections of the floors could lead to many issues such as the amplification of higher modes of vibration and increased structural damage. For these reasons it is suggested to design the floors as non-ductile elements, with overstrength connections [18].

3.3. Role of the Mechanical Connections

Mechanical connections in CLT structures can be divided in two groups. The first refers to connections selected to prevent rocking (HD) and translation (AB) of the vertical panels. The second includes small diameter metal fasteners used for vertical joints in segmented walls, or for panel to panel connections of the floors, or for floor to wall connections.

To date, only HD and AB placed in panel to panel and in panel to foundation connection zones—and vertical joints in segmented walls—are considered as potential dissipative fuses against seismic actions. Experimental tests highlighted that in the majority of cases ductile failure modes defined by the Johansen's theory [39] have occurred for HD and AB loaded in tension and in shear, respectively. These mechanisms are characterized by a combination of progressive embedding of the wood and cyclic yielding of the steel fasteners (nails or screws) with a good hysteretic behaviour.

Extensive programs aimed at investigating the cyclic behaviour of HD and AB have been conducted, for instance within the SOFIE [21] and SERIES [40] projects. Gavric et al. [21] investigated cyclic behaviour of HD and AB under different loads configuration: HD subjected to tension, HD subjected to shear, AB subjected to shear and AB subjected to tension. Test results demonstrated that HD tested in tension show high strength and stiffness in their primary direction (tension), while in shear direction they do not achieve high strength and stiffness due to local buckling of steel plates. AB, instead, have significant strength and stiffness in both directions. Cyclic response of HD is characterized by wide loops in tension and null in compression, while AB have a hysteretic response with pinching due to the slip of the fasteners in the embedded holes of the wood. In Figure 7 are reported the force-displacement diagrams relative to HD and AB loaded in tension and in shear, respectively. Other similar experimental campaigns finalized to evaluate the mechanical behaviour of connections used in low and mid-rise CLT buildings have been conducted in [41–43], in which the same results have been achieved.

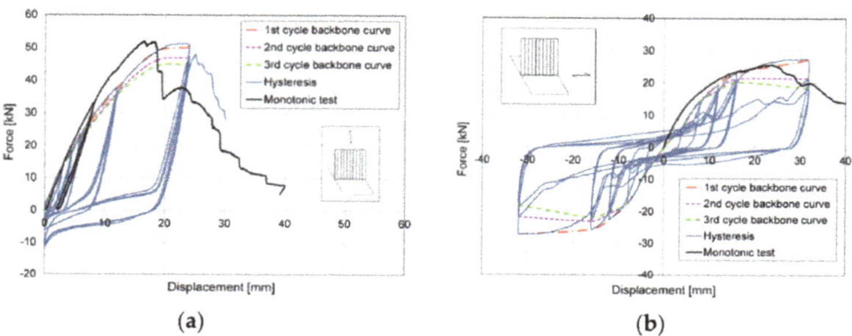

Figure 7. Typical experimental cyclic force-displacement behaviour of traditional connections: (**a**) hold-down loaded in tension, (**b**) angle-bracket loaded in shear (from [19]).

Although in design practice it is commonly assumed that AB carry-out shear forces and HD tension forces only, a simultaneous presence of shear–axial interaction occurs during an earthquake. The coupled shear-to-tension behaviour for HD [44] and AB [45] have been also investigated. Results proved that when HD and AB connections are subjected to significant combined shear and axial loads, the global ductility and dissipative capacity of the entire wall panels assembly can be significantly deteriorated.

4. Modeling Strategies of the Components

4.1. Material Modeling

In solid mechanics, CLT panels have a complex anisotropic behaviour. For structural applications (i.e., at macroscale level) the panels can be schematized as an orthotropic multi-layer material with different in-plane and out-of-plane behaviour.

Due to its strength and stiffness, the in-plane displacement capacity of shear-walls depends almost exclusively on the deformability induced by the mechanical connections, thus the CLT panel is often supposed to behave as an infinitely rigid body. On the contrary, in the case of floors the transverse layers induce significant shear deformations and influence significantly their out-of-plane flexural deflections and vibrations under service loads. Such differences of behaviour between in-plane and out-of-plane require differentiating the structural modelling of the panels in the two planes.

4.1.1. In-Plane Modeling

The cross-layered morphology of CLT panels and the presence of an uneven number of layers involve higher strength and stiffness in the longitudinal direction than in the transversal one. Experimental tests and numerical analyses proved that the in-plane displacement capacity of single or multi-story CLT shear-walls is dominated by rocking and translation contributions due to connections, while the own bending and shear deformabilities of the material are negligible because they influence by less than 5% the total plastic horizontal displacement at the top of the panel. [46,47]. Anyhow, a refined model to estimate the in-plane deflections for single and coupled CLT walls under horizontal loadings is presented in Shahnewaz et al. [37], in which also the influence of orthogonal walls and floors has been taken into account.

Macroscale models used in literature to schematize the CLT shear-walls are the homogenized orthotropic material (HOM) or the homogenized isotropic material (HIM), and their employment depends on the possibilities offered by the used finite element software. The first approach, more sophisticated, simulates the elastic properties of material (Young's, tangential and Poisson's moduli) in the longitudinal and transverse layers based on the theory of orthotropic membrane and is suitable for studies aimed at investigating the local behavior of the material.

For design practice, i.e., when the influence of material deformability is not of paramount importance, Blass and Fellomoser [48] proposed adopting HIM simplified model. It consists of reduction of a multi-layer to a single-layer section by means of the coefficient k_3 and k_4 (as a function of longitudinal or transversal direction of the panel) listed in Table 2. A further simplification of the HIM method has been introduced in [49,50] that consists of reducing Young's modulus of raw material as a function of the effective number of layers (Table 2). Moreover, in the second method also tangential elastic modulus is suggested to reduces as a function of the effective number of layers, as for Young's modulus. When the lamellae of the panels are not glued edge to edge a further reduction of 10% of the Young's modulus is advisable, as suggested in [20].

Table 2. Coefficients for effective stiffness for cross-lam panels loaded in-plane.

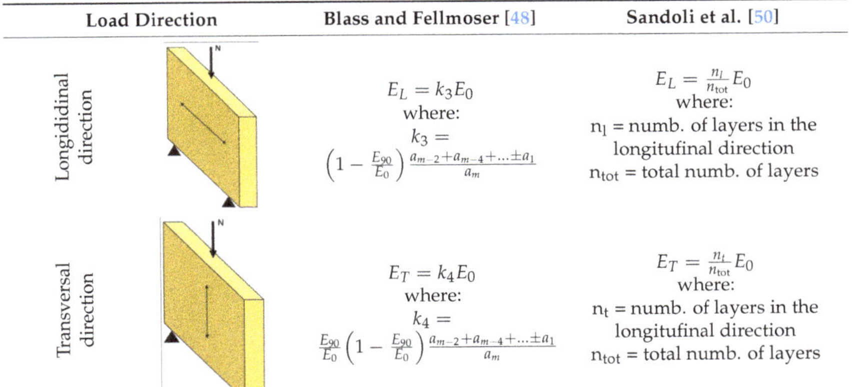

Load Direction		Blass and Fellmoser [48]	Sandoli et al. [50]
Longididinal direction		$E_L = k_3 E_0$ where: $k_3 =$ $\left(1 - \frac{E_{90}}{E_0}\right)\frac{a_{m-2}+a_{m-4}+\dots\pm a_1}{a_m}$	$E_L = \frac{n_l}{n_{tot}} E_0$ where: n_l = numb. of layers in the longitufinal direction n_{tot} = total numb. of layers
Transversal direction		$E_T = k_4 E_0$ where: $k_4 =$ $\frac{E_{90}}{E_0}\left(1 - \frac{E_{90}}{E_0}\right)\frac{a_{m-2}+a_{m-4}+\dots\pm a_1}{a_m}$	$E_T = \frac{n_t}{n_{tot}} E_0$ where: n_t = numb. of layers in the longitufinal direction n_{tot} = total numb. of layers

a_m = panel total thickness, a_1 = thickness of the middle layer, a_{m-2} = distance between two transverse layers (measured between the center of gravity of each layer), E_0 = Young's modulus of raw material.

4.1.2. Out-of-Plane Modeling

The out-of-plane flexural behaviour of CLT panels is significantly influenced by the shear deformations of the transverse layers. This phenomenon, indicated as "rolling shear", is due to the tangential stresses (τ) acting in the orthogonal to grain direction within transverse layers of the panels. Due to the low value of the shear or tangential modulus in the orthogonal to grain direction of timber boards (named rolling shear modulus $-G_R$), grains rolling off one from each other occur in transverse layers (Figure 8). As a consequence, increases of deflections and vibrations of the panel and modification of its internal stresses arise [51,52].

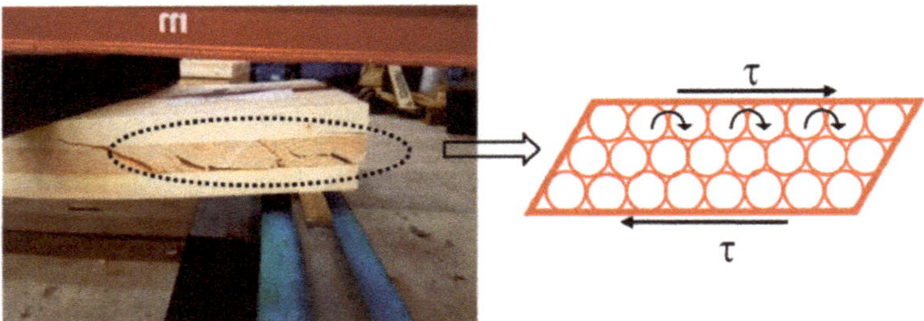

Figure 8. The "rolling shear" effect in CLT panels subjected to out-of-plane flexure.

The rolling shear phenomenon was first raised by Blass and Fellmoser [51], who highlighted its effect on the flexural behaviour of CLT floors having different length-to-depth aspect ratios. Experimental studies aimed at investigating different aspects have been conducted [53–56], as well as numerical investigations. A multi-scale finite element model accounting for the rolling shear failure by means of an homogenization and cohesive zone model of the panel has been proposed by Saavedra Flores et al. [57]; such a model is able to capture the interlaminar and inter-fibre cracking and to solve the macroscopic equilibrium using computations. Strurzenbecher et al. [58] investigated the structural design and modelling by advanced plates theory which combines accuracy of the description of deformations and stresses in the plate with computational efficiency, and compared the results with analytical solutions.

Consequently, practical tools for engineering applications aimed at evaluating deflections affected by the rolling shear have been proposed in literature, and part of them has been included in manuals or guidelines [2,11]. These models are based on the beam theory, but take into account shear deformations in transverse layers. For instance the "γ-method" (also included in Eurocode 5) allows consideration for composite beams joined by metal fasteners, where the transverse cross-layers in CLT are considered as equivalent shear fasteners uniformly distributed over the length of the panel. In Table 3 are reported the formula for evaluating the effective flexural stiffness $(EI)_{\text{eff}}$ of the panels through the coefficient γ which takes into account the shear deformations in the transverse layers.

Table 3. Analytical formulas to evaluate the out-of-plane stiffness of CLT panels.

Stiffness-Type	γ-Method	Shear Analogy Meth.
Flexural stiffness	$(EI)_{eff} = \sum_i (EI)_i + \sum_i (\gamma E A a^2)_i$ where: $\gamma = \left[1 + \frac{\pi^2 (EA\Delta z)}{(kl^2)} \right]^{-1}$ $k = \frac{G_R\,B}{h} \Delta z$	$(EI)_{eff} = (EI)_A + (EI)_B$ where: $(EI)_A = \sum_i (EI)_i$ $(EI)_B = \sum_i (EAa^2)_i$
Shear stiffness	is taken into account trought the coeffieicent γ	$(GA)_B = \dfrac{Ba^2}{\left[\frac{h_1}{2G_1} + \sum_{i=2}^{n-1} \frac{h_i}{G_i} + \frac{h_n}{2G_n} \right]}$

E = Young's Modulus, G = shear modulus, G_R = rolling shear modulus, I = moment of inertia, A = cross section area, a = distance between the centre of gravity of the single layer and the total one, k = equivalent stiffness provided by transverse layers, l = panel length, B = panel width, h = panel height.

Another method is the "shear analogy method" introduced by Kreuzinger [59]. Therein, the effective flexural stiffness $(EI)_{\text{eff}}$ and the shear stiffness (GA) are assigned to two displacement-coupled beams: Beam A is given by the sum of the inherent flexural and shear stiffness of the individual layer along its own centres, while beam B is given by the Huygens–Stainer points, or an increased moment of inertia because of the distance from the neutral axis of the flexural and shear stiffness of the panel. These beams are coupled with infinitely rigid web elements in order to ensure the same elastic curvature between beams A and B. In Table 3 are reported the formulas for calculating flexural and shear stiffness trough the "shear analogy method".

A comparative analysis between these two methods, carried out on panels with length-to-depth rations from 5 to 50, has been recently presented in [52]. The authors concluded that for ratios ranging from 25 to 40 (typically used in CLT buildings) the results given by the "γ-method" and the "shear analogy method" are comparable, either in terms of deflections and internal stress state of the panels.

4.2. Connection Modeling

As mentioned, in design practice the in-plane flexural behavior of CLT shear-walls is ruled by rocking and translation due to connections. Thus, to evaluate the flexural load-bearing capacity of the panel, an adequate modeling of the connections is of paramount importance.

The majority of methods presented in literature assume that the panel to panel and the panel to foundation are modelled in analogy with reinforced concrete cross-sections. The HDs represent the tensile-resistant elements (e.g., steel bars), ABs are the shear-resistant elements (e.g., stirrups), while timber to timber contact represents the concrete in the compressed zone (this latter is defined by a neutral axis depth).

Two different groups of methods can be defined: (a) simplified and (b) detailed. In the first case the contribution of timber in the compressed zone is disregarded, while in the second case it is considered. A simplified mechanical model to evaluate the in-plane flexural load-carrying capacity of CLT panels has been proposed in [60]. It is based on the equilibrium between the internal forces developed by HD and the overturning moment, assuming that the panel rotates around its edge and that the tensile stress distribution at the base cross-section of the panel is linear (Figure 9a). Gavric and Popovski [61], instead, refined the methodology proposed in [60] considering also the possibility of the shear to tension interaction in the connectors, and in particular in the ABs as tests showed that HDs do not provide significant shear resistance.

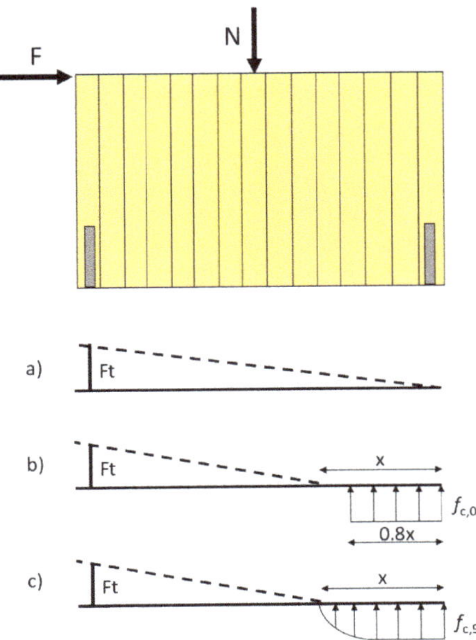

Figure 9. Assumption on normal stress distributions at a pane to panel and panel to foundation contact.

Pei et al. [62], introduced another simplified method based on a linear tensile stress distribution again, but the connector's elongation and its stiffness and strength are considered. The tensile strength of each connector is proportional to the distance of the connector from the panel edge. A triangular distribution of the connector displacement is considered based on the fact that the furthest connector reaches its total elastic tensile strength.

The timber to timber or timber to foundation compression strength in evaluating the load-carrying capacity of CLT panels has been considered in [63]. Herein, the compressed zone is modelled by means of a stress-block diagram with maximum design strength equal to that of timber in parallel to grain direction ($f_{c,0,d}$) (Figure 9b). An advancement of this method is presented in [64], where both elongation capacity of connection elements and the compressed timber contribution are considered. In particular, the maximum compression strength in parallel to grain direction and the Young's (E_{90}) modulus in orthogonal to grain direction have been adopted for the compressed timber in the case of timber to timber contact; while for timber to (reinforced concrete) foundation contact the Young's modulus (E_0) in parallel to grain direction has been assumed.

A further study recently developed in [50] proposes a theoretical model which includes both orthogonal to grain compression strength ($f_{c,90}$) and orthogonal to grain Young' modulus for timber to timber contact (while E_{90} and $f_{c,90}$ are changed with E_0 and $f_{c,0}$ in the case of timber to concrete contact at the foundation) (Figure 9c). A parabola-rectangular stress ($\sigma_{c,90}$)-strain ($\varepsilon_{c,90}$) model has been proposed (using the relationships proposed by Eurocode 2 for the concrete) to model the compressed timber in orthogonal to grain direction.

Such methods remain also valid in the case of coupled walls, provided that the contribution of the vertical joints is included in the evaluation of the flexural strength of the panels. Simplified model which do not consider the compressed timber contribution can be found in [47], while in [49] the panel to panel orthogonal to grain compression stresses have been also included in the model.

In a recent study, Franco et al. [65] introduced the effect of shear to axial force interaction on the flexural capacity of CLT panels. They adopted an elastic-perfectly plastic constitutive law for timber and coupling effect between the axial and lateral strength of the connections, allowing for a more realistic representation of the flexural behaviour of the panels.

5. Modeling of Multi-Story Timber Buildings

The adoption of finite element numerical models for multi-storey CLT buildings became of particular importance in both the research field and design practice. The execution of full or reduced-scale tests is expansive and time consusming procedure because of the several parameters that affect the seismic behavior of CLT shear-walls. Thus, the availability of reliable numerical models makes it fundamental for engineers to perform both linear and non-linear seimic analyses. With this aim, different finite element methods, more or less simplified, for multi-story CLT buildings presented in literaure are discussed in the following sections.

5.1. Vertical Walls

The in-plane behavior of CLT multi-story shear-walls is generally modelled through 2D-shell elements, whose material elastic properties are calculated with one of the methods described in Section 4.1.1. Nevertheless, alternative approaches are available in literature: for instance a set of trusses (pinned in the corners) with a diagonal truss or springs simulating the in-plane strength and stiffness of the panels [66,67]. With respect to 2D-shell models, truss elements cannot make visible the whole stress path in the panel but only the resultant of the internal forces in the elements, resulting in being less effective in the case of walls with openings. In fact, it has been proved that, in the case of perforated walls, a stress concentration in the corners at the interfaces between lintels and vertical panels can arise. Such stresses can be higher than the design strength values of material and lead to local brittle failure of the material in the corner region due to tension or compression [23,50,68].

Furthermore, also in case of segmented walls having lintels connected by means of metal fasteners to the adjacent vertical panels, brittle failure modes due to stresses concentration in the corners can precede the ductile mechanisms of the connections [20].

Another sophisticated model based on the use of 3D-solid elements which considers both thickness of the panels and layer orientation of the boards has been presented in [69].

Particular attention should be paid in modeling of perforated walls, because they are affected by the lintels. As discussed in Section 3.1, the lintels may have a different lay-out depending on how the wall is conceived, as a monolithic or segmented. In the monolithic case, the number of effective vertical layers against flexure and shear forces is less than that in the vertical panels (Figure 10a). On the contrary, in the case of segmented walls the lintels can have the same number of effective layers of the vertical panels (Figure 10 b). In this latter case the lintels can be (i) less or more effectively fastened to the adjacent vertical panels by means of metallic fasteners or (ii) simply supported on adjacent vertical panels (without fasteners).

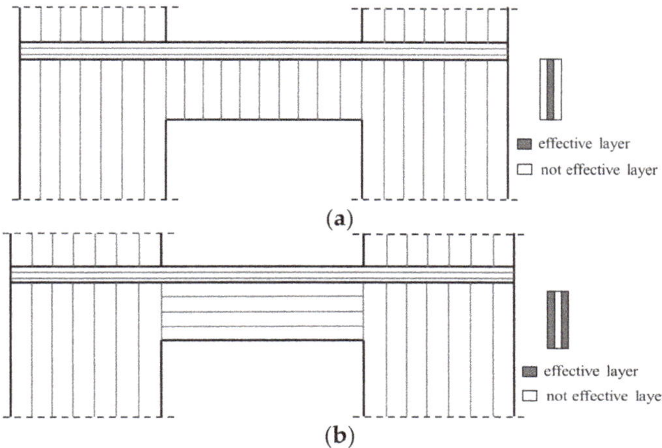

Figure 10. Lintels configurations in case of: (**a**) monolitich wall, (**b**) segmented walls.

The seismic behavior of CLT walls with opening is also affected by the relative stiffness of the mechanical connections at the base of the panels with respect to the in-plane stiffness of the wall (and the lintels). Such an issue has been investigated in Mestar et al. [70], where a kinematic model to analyze the effect of different hold-down configurations on the overall in-plane stiffness has been developed, and the results obtained compared with the experimental ones. They concluded that the degree of coupling decreases with increased hold-down stiffness and increases with wall segment width.

5.2. Floors

As well as for the shear-walls, the in-plane behavior of CLT floors can be schematized with 2D-shell elements having material properties calculated as descibed in Section 4.1.2, or through truss elements having diagonals (made with trusses or springs) to simulate the in-plane strength and stiffness of the floors [71,72].

Lateral force transferring a diaphragm represents a topic still under study by reserchers, then few indications can be found in both research papers and codes. For practical purposes, CLT floors are assumed as rigid diaphragms in the finite element model of the buildings: they are capable to transfer horizontal seismic actions to the vertical walls, provided that panel-to-panel connections are suffciently stiff and resistant.

The assumption of a rigid diaphgram is of particular importance in seismic analyses, because flexibility of floors can influence the fundamental vibration modes of the structures changing the force transmission on the vertical elements [71]. When such an assumption is made, NDS [33] suggests verifying it by calculating the maximum in-plane deflection of the floors. For a simple supported floor, having length L and width W, maximum deflection can be deermined as in the following (Figure 11):

$$\delta = \frac{5vL^3}{8EAW} + \frac{wL^2}{8A_{clt}G_{ef}} + CLe_n + \frac{\sum(x\Delta_c)}{2W} \tag{1}$$

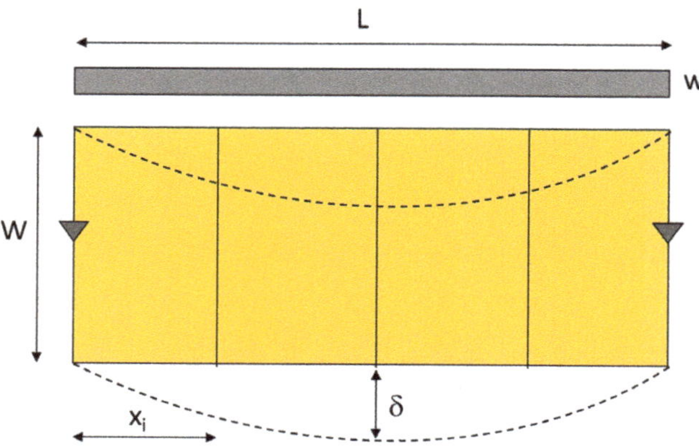

Figure 11. In-plane deflection of CLT floors under seismic actions.

The first term of the Equation (1) represents the flexural deflection contribution, where $v = wL/2B$ is the maximum shear at the edge of the diaphgram calculated from the uniformly distributed load w, B is the diaphragm width, E is the Young modulus and A the cross-section area of the effective lamination layers. The second term is the shear deformation of the CLT panels, where A_{clt} is is the gross cross-sectional area of the CLT diaphragm resisting the shear load and G_{ef} is the effective shear modulus; this latter can be estimated as in Flaig and Blass 2013 [73]. The third term takes into account the slip contribution of the connection on the total deflection: the value of C is calculated as $C = 0.5(1/P_L + 1/P_W)$ where P_L and P_W represent the length and the width of the typical panel respectively, while e_n is the connection slip at the outer edge of the diaphgram under the maximum shear load v. The derivation of e_n can be found in the ATC-7 [74], derived for application to typical wood structural panels diaphgrams. The fourth term is the geometric translation of any chord slip under loading into lateral deformation of the diaphragm, where $\Delta_C = 2(\text{T or } C)/\gamma n$ (T and C are the maximum tensile and compression forces due to applied loads w, γ is the load slip modulus and n the number of fasteners) and x is the length considered for evaluating maximum deflection due to chord slip.

By contrast, when the floors are modelled by means of 2D-shell elements or trusses the maximum in-plane deflection can be determined as a result of the numerical analyses.

As regards the schematization of the out-of-plane behavior, it is limited to evaluate the amount of the out-of-plane flexural deflections or vibrations. Literature papers demonstrated that the analytical model described in the Section 4.1.2 gives back reliable results to calculate flexural deflections, vibrations and stress state, comparable with those obtained with finite element models composed of multi-layer material [52]. In any case, the adoption of the multi-layer material surely gives back more consistent results and allows conducting modal analyses to calculate accurately maximum vibrations.

5.3. Mechanical Connections

In multi-story CLT buildings the mechanical connections are usually modelled through links, trusses or frame elements.

According to literature, uniaxial or biaxial behavior is generally considered to simulate the connections [23,46,50,75–77]. In the first case, HDs and ABs resist only in their primary direction (e.g., HDs only in tension and ABs only in shear), in the second case both HDs and ABs resist axial and shear forces simultaneously. Uniaxial behavior is commonly used in the practical applications and leads to more conservative results with respect to the biaxial behavior. Moreover, for detailed analyses (e.g., time history analyses) the tension to shear force interaction domains are available to model the biaxial behaviour [61].

The models discussed in Section 4.2 are used to schematize the panel to panel and the panel to foundation connection zones in finite element models of multi-storey CLT buildings. Both 'simplified' or 'detailed' approaches discussed can be used to model the connection regions. Simplified methods do not consider the timber to timber contact contribution, then the connection model consists of: (a) HD truss or spring elements, which simulate the tensile behaviour of the hold-downs devoted to face tensile forces due to rocking of the panel; (b) AB truss or spring elements, which simulate the angle brackets devoted to resist shear forces due to translation of the panel. In such a model HD and AB are modelled as no-compression elements (due to buckling phenomena of the stem of HDs not interested by fasteners, when subjected to compression forces).

To obtain more realistic analytical results, the contribution of the compressed timber should be included in the structural model. Thus, additional no-tension truss or spring elements devoted to simulate the timber to timber contact have been introduced in the structural model developed in references [50,64,75].

Three different approaches to model the non-linear behaviour of the connections, and considered as representative for this paper's purpose, are reported in Table 4. An effective but sophisticated approach to model the cyclic behaviour of the connection has been presented by Rinaldin et al. [77]. The authors proposed the adoption of multi-linear hysteretic cyclic behaviour for both HDs and ABs, implemented thought a multilinear spring placed at the base of finite element model of the wall. Such schematization allows dynamic non-linear analyses to be carried out. Instead, a simplified monotonic trilinear model implemented within non-linear springs has been suggested by Yasumura et al. [63] to carry out non-linear static analyses. Simplified elastic-perfectly plastic force-displacement behaviours have been also proposed in Vassallo et al. [76] and Sandoli et al. [50]. The first difference between these two latter models is that in Sandoli et al. [50] the timber to timber contact takes into account the orthogonal to grain timber properties, while in Vassallo et al. [76] only that in parallel to grain direction is considered. The second difference is that in Vassallo et al. [76] the shear forces are faced through a couple of diagonal trusses joined at the corner of the panel, while in Sandoli et al. [50] through frame elements.

Table 4. Literature models for connections.

Authors	Structural Model	Connection Model
	Trilinear Cyclic Behavior	
Rinaldin et al. [77]	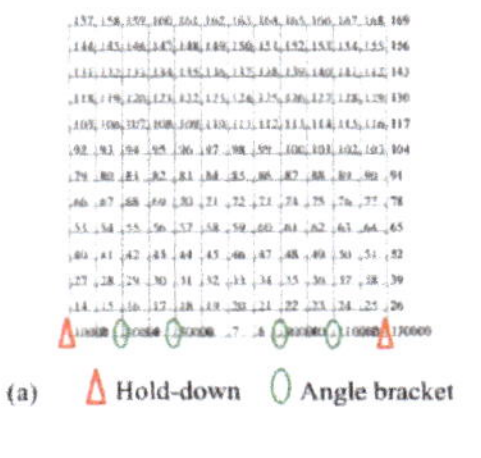 (a) △ Hold-down ○ Angle bracket	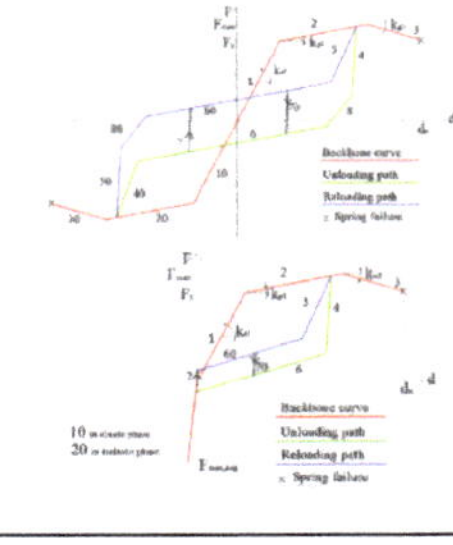 Backbone curve / Unloading path / Reloading path / Spring failure

Table 4. *Cont.*

Authors	Structural Model	Connection Model
	Trilinear Monothonic Behavior	
Yasumura et al. [68]		
	Bilinear Monotonic Behaviour	
Vassallo et al. [77]		Hold-down (HD) & angle brackets (AB)
Sandoli et al. [50]		Hold-down (HD) & angle brackets (AB) / Timber-to-timber contact (C)

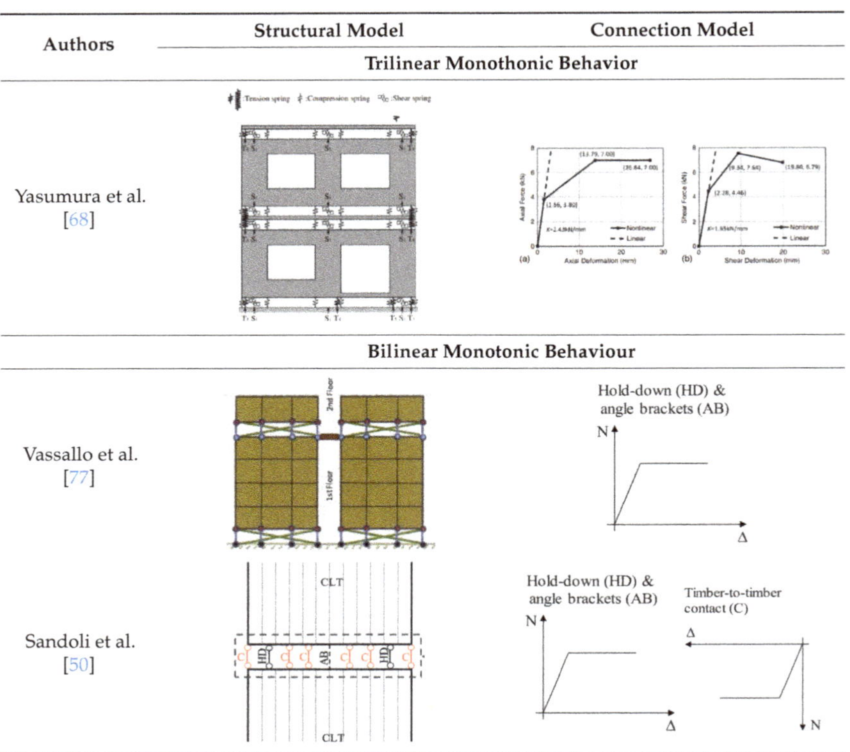

These connection models are also valid to schematize the panel to panel and the panel to foundation connection zones in the case of coupled (or segmented) CLT wall, provided that additional shear springs to simulate the shear stiffness and strength of the fasteners in the vertical joints between the panels are included in the structural model [47,49,78].

6. Future Trends in CLT Building Practices

6.1. Damage of the Timber Components

The modern performance-based seismic design (PBD) method entails high-performant and high-safety structures, capable of overcoming severe earthquakes exploiting all their available strength, stiffness and ductility. Nevertheless, the reduction of (i) damaging of structural and non-structural components, (ii) economic losses after the seismic event and of the (iii) environmental impact during the entire life of the construction (including the disposal and recycling) are also required by PBD.

Experimental tests on connections highlighted serious damage in panel to panel and in panel to foundation connection zones: in spite of a ductile response of the whole connections, vertical timber panels showed diffused permanent damage after the tests due to timber embedding produced by the interaction between timber boards and metal fasteners (Figure 12), while metal connectors yield.

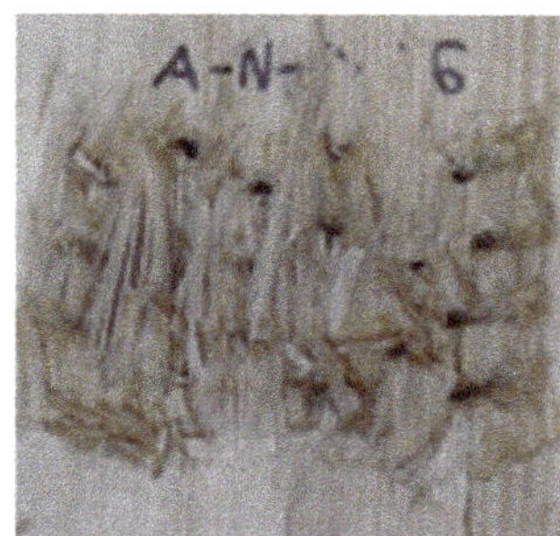

Figure 12. Damage to CLT vertical panels due to embedding.

It should be pointed out that the current design method proposed in manuals, literature papers or guidelines provided that Johansen's ductile failure modes develop in the mechanical connections (HD and AB) [11,15,31,68,75]. Some of these mechanisms are reported in Figure 13a–c. Sometimes ductile mechanisms have been anticipated by brittle failure modes caused (i) by the reduced end edged spacing or underestimation of the actual strength of the ductile components, with consequent increased strength demand for the brittle parts, which may fail if designed with insufficient overstrength (Figure 13d,e) [25,26]; or (ii) by high number of metal fasteners which produce the so called "group effect" (Figure 13f) [11].

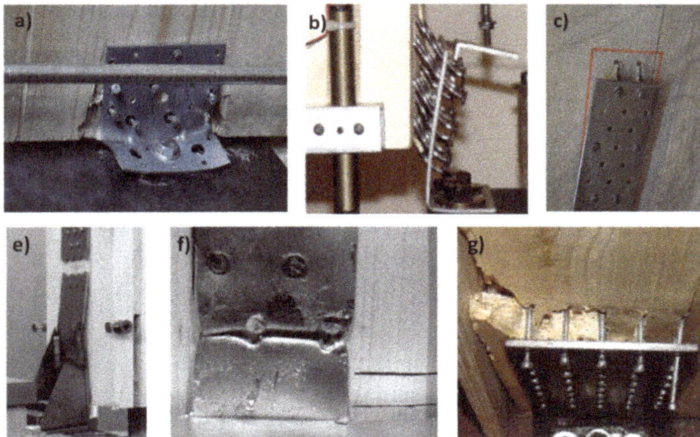

Figure 13. Typical failure modes of the traditional connections: (**a**) Gavric et al. [21], (**b,c**) Schneider et al. [79], (**e–g**). FP innovation handbook [11].

Subjected to seismic actions, also the horizontal panels (floors) will remain permanently damaged. The compression forces transmitted by the rocking motion of the vertical CLT panels involve an irreversible crushing of the floor panels in the timber to timber contact zones [50]. In particular, the compression forces load the floor panels in orthogonal to grains direction, where low strength and stiffness of material is expected ($f_{c,0}/f_{c,90} \cong 10$, $E_0/E_{90} \cong 30$).

Therefore, both vertical and horizontal panels will be unusable after a seismic event, as well as the entire construction, with considerable economic losses and environmental impact as replacing materials timber and steel increases the carbon dioxide release in the atmosphere.

In the last few years, "integrated design" approach between seismic performance and environmental sustainability attracted researchers' attention. The key concept is that sustainability means not only ensuring the best heating and cooling energy consumption of the buildings with respect to the seasonal cycles but also guaranteeing optimal seismic behaviour with regard to damage reduction in the structural (and non-structural) elements.

6.2. New Frontiers for More Sustainable CLT Structures

It is recognized that CLT buildings are provided with high environmental sustainability during their life cycle, with respect to other structural types (e.g., reinforced concrete). This depends, firstly, on the excellent eco-sustainability attitude of raw material and, secondly, on the attention paid by the manufacturers to produce certified environmentally sustainable materials.

The sustainability of CLT buildings can be furtherly enhanced within the life cycle of the construction if the amount of material disposal and/or its recycling is reduced. Seismic damage to timber parts is one of the main causes of material disposal and recycle, and it appears fundamental to mitigate seismic damage in the connection zones.

In order to reduce damage in timber panels, some alternative connection systems have been recently introduced in literature. The common goals of such systems are those of:

- concentrating the hysteretic behaviour in metal parts (i.e., steel plates or similar devices) only;
- eliminating possibilities of damaging in timber components (both timber embedding of CLT vertical panels and orthogonal to grain crushing of horizontal ones).

To satisfy these requirements, a new way of conceiving the connection zones is needed. The current hierarchy of strength criterion based on the concept of the strength of metal plates–weak timber to steel connections (i.e., timber to fasteners interaction) should be inverted as strong CLT panels–weak metal devices. To obtain this, cyclic dissipation must be concentrated in steel devices properly conceived as regards this aim.

Such solutions are recognized as low-damage connections. They represent an innovative aspect in timber buildings practice, even if they arose in the United States in the 1990s for post-tensioned precast concrete structures [80] and subsequently were adapted in New Zealand to post-tensioned rocking timber structures [81,82]. Dissipative connections proposed in New Zealand consist of fuse-type elements made with mild steel (or viscous damping) dissipaters placed at the base of the panels, replaceable after seismic damaging (reason for which are also named "plug and play" systems) (Figure 14).

Figure 14. Plug and play fuse-type external dissipaters (from [82]).

A low-damage philosophy can be also applied to the traditional CLT buildings, adopting connection systems devoted to hysteretic energy dissipation similar to those used for post-tensioned structures. The main objective is that under seismic actions the vertical timber panels remain elastic and undamaged, thus they can be reused to attach new connections after seismic damage. With respect to traditional HD and AB, the low-damage connections rely on the dissipation of steel part only, with further advantages: cyclic behaviour presents wide loops without pinching and strength and stiffness degradation is that of steel.

Replaceable connections can be made with mild-steel bars, or thin steel plates nailed or screwed to timber panels having a properly designed shape. For instance, Latour and Rizzano [25] proposed a new type of hold-down called the XL-stud, where the main idea is to shift the dissipative zone of the hold-down from the stem zone to the flange plate (Figure 15a). The stem zone has to be over-strengthened by adopting a proper number of nails and by checking the resistance of stem and wood with respect to the force needed to yield the flange plate. Moreover, to maximize the plate energy dissipation, the proposal is to give an hourglass shape to the flange plate of the angle (Figure 15b).

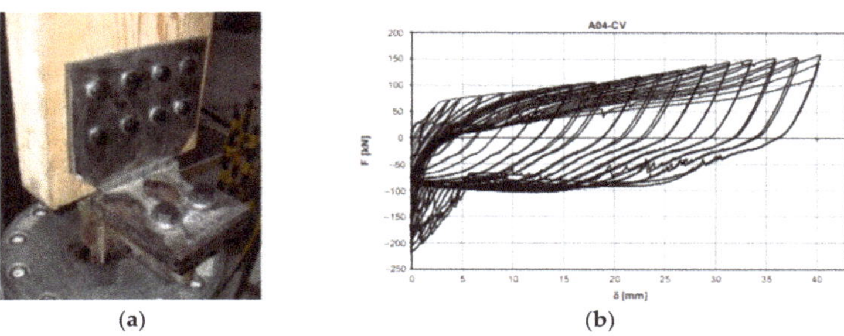

(a) (b)

Figure 15. XL-stud connection system: (**a**) XL-stud shape, (**b**) cyclic behaviour (from [25]).

Scotta et al. [26] proposed another innovative dissipative connection (named X-bracket) composed of X-shaped metallic plates nailed or screwed to timber panels, used to withstand either tensile and shear forces in panel to panel, panel to foundation and wall to wall connections (Figure 16a,b respectively). Energy dissipation involves the steel plates only, while timber parts are overstrength. The X-shape is optimized to prevent localized failures and to assure diffuse yielding of material, emphasizing ductility and energy dissipation capacity (Figure 16c,d).

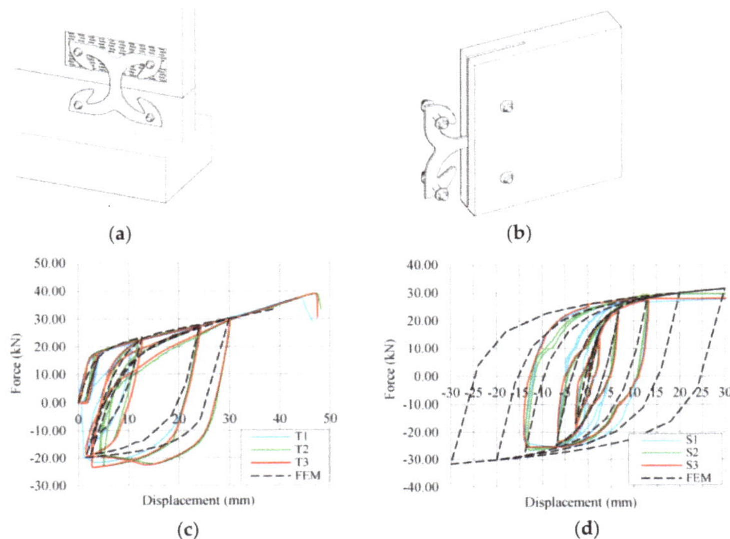

(a) (b)

(c) (d)

Figure 16. X-brackets connector: (**a**) panel to foundation connection, (**b**) wall to wall connection, (**c**) tensile test, (**d**) shear test (from [26]).

The issue of damage to timber parts has been also raised in Sandoli et al. [50]. They proposed an alternative dog bone-shaped HD to perform predictive nonlinear static analyses, whilst experimental results are not still presented. As for Latour and Rizzano [25], the energy dissipation is concentrated in the stem of the HD not interested by fasteners, such that the timber-to-steel interaction remains elastic and CLT panel undamaged.

To date, there is no experimental evidence in the literature of orthogonal to grain crushing of horizontal floors due to compression forces transmitted by rocking motion of the vertical panels, nor has it been observed on real buildings hit by earthquakes. On the other hand it is evident that such damage is related to the platform constructional technology currently adopted for CLT constructions, which requires floors interposed between two consecutive vertical panels. As an alternative, a balloon-type technique permits such a problem to be overcome, because the floors are connected on the internal side of the vertical walls by means of specific mechanical connections. The balloon technique has been already used in building practice in New Zealand for low-damage timber constructions having CLT or LVL shear-walls as seismic-resistant system and timber frames to resist gravity loads. Moreover, it is comforting that studies on mechanical behavior of balloon CLT shear-walls are ongoing with regard to CLT buildings [38].

7. Conclusions and Outlooks

This paper reports a state-of-the art overview on seismic performance and sustainability aspects of CLT buildings in seismic-prone regions. In general, the growing interest of traditional CLT buildings, i.e., realized with a platform construction technique and connection made with hold-downs and angle-brackets, for the residential and non-residential construction market has been remarked upon in the paper.

Such development is mainly due to the high sustainability of material on the whole life cycle (from extraction of raw material, to production and usage, up to disposal and recycle). Despite the fact that no observations on the behavior of seismically-designed CLT buildings hit by real earthquakes are available to date, experimental and numerical studies highlighted the particular attitude of CLT buildings to withstand severe earthquakes thanks to lightness of material and to significant dissipation capacity of the connections. Numerous experimental tests on traditional connections (or subassemblies) and modelling methods for multi-story CLT buildings subjected to seismic actions have been proposed in literature. Such methods, giving reliability and easy application to engineers, are still not reflected in national or international design codes.

The state-of-the-art overview allowed us also to draw conclusions on future trends in buildings practice aimed at enhancing both seismic performance and environmental sustainability.

Recognizing the basic attribute of CLT buildings to exhibit good seismic performance, it has been pointed out as the reduction of structural damage in timber components can further enhance the already high environmental sustainability in the phases of disposal and/or recycling of material. More than with other structural types, CLT buildings are particularly suitable to changes of technological and constructional details because they are characterized by the level of prefabrication. Some researchers are studying alternative (low-damage) connection systems to prevent seismic damage to vertical timber walls in the case of moderate and severe earthquakes, or alternative constructional details (balloon-type construction instead of platform-type) to prevent permanent damage to floor panels in timber to timber contact regions. The goals of such systems are, respectively, those of concentrating the hysteretic behavior in metal parts only (i.e., steel plates or similar devices) and of eliminate orthogonal to grain permanent crushing of panel floors due to the rocking motion of the vertical panels.

Although in the majority of cases only experimental results are available and any conclusive consideration can be made to date, alternative low-damage solutions seem to be reasonably effective for damage reduction in timber elements. Future advancements will be necessary, devoted to validating such systems with regard to technological feasibility, numerical modelling and the effectiveness of life cycle assessment.

Author Contributions: Conceptualization, A.S. and B.C.; methodology, A.S., C.C., C.D.; validation, B.C. and A.P.; data curation, A.S. and B.C.; writing—original draft preparation, A.S.; writing—review and editing, A.S., C.C., C.D. and A.P.; visualization, C.C. and B.C.; supervision, B.C. and A.P. All authors have read and agreed to the published version of the manuscript.

Funding: This research received no external funding.

Conflicts of Interest: The authors declare no conflict of interest.

References

1. Brandner, R.; Flatscher, G.; Ringhofer, A.; Schickhofer, G.; Thiel, A. Cross laminated timber (CLT): Overview and development. *Eur. J. Wood Prod.* **2016**, *74*, 331–351. [CrossRef]
2. Swedish Wood. *The CLT Handbook-CLT Structures, Fact and Planning*; Swedish Wood: Stockholm, Sweden, 2019.
3. Iqbal, A. Cross Laminated Timber in New Zealand: Introduction, prospects and challenges. *N. Zealand Timb. Design J.* **2018**, *22*, 3–8.
4. Pei, S.; Rammer, D.; Popovski, M.; Williamson, T.; Line, P.; van de Lindt, J.W. An overview of CLT research and implementation in North America. In Proceedings of the WCTE, Vienna, Austria, 22–25 August 2016.
5. Goto, Y.; Jockwer, R.; Kobayashi, K.; Karube, Y.; Fukuyama, H. Legislative background and building culture for the design of timber structures in Europe and Japan. In Proceedings of the WCTE, Seoul, Korea, 20–23 August 2018.
6. Chen, C.X.; Pierobon, F.; Ganguly, I. Life Cycle Assessment (LCA) of Cross-Laminated Timber (CLT) Produced in Western Washington: The Role of Logistics and Wood Species Mix. *Sustainability* **2019**, *11*, 1278. [CrossRef]
7. Borjesson, P.; Gustavsson, L. Greenhouse gas balances in building construction: Wood versus concrete from life-cycle and forest land-use perspectives. *Energy Policy* **2000**, *28*, 575–588. [CrossRef]
8. Guo, H.; Liu, Y.; Meng, Y.; Huang, H.; Sun, C.; Shao, Y. A Comparison of the Energy Saving and Carbon Reduction Performance between Reinforced Concrete and Cross-Laminated Timber Structures in Residential Buildings in the Severe Cold Region of China. *Sustainability* **2017**, *9*, 1426. [CrossRef]
9. Liu, Y.; Guo, H.; Sun, C.; Chang, W.-S. Assessing cross laminated timber (CLT) as an alternative material for mid-rise residential buildings in cold regions in China—A life-cycle assessment approach. *Sustainability* **2016**, *8*, 1047. [CrossRef]
10. Suzuki, M.; Oka, T.; Okada, K. The estimation of energy consumption and CO_2 emission due to housing construction in Japan. *Energy Build.* **1995**, *22*, 165–169. [CrossRef]
11. *CLT Handbook: Cross Laminated Timber; FP Innovation Special Publication SP-529E*; U.S. Edition: St. Jean Pointe-Claire, QC, Canada, 2013.
12. EN 1995-1-1: 2004. *Eurocode 5: Design of Timber Structures—Part-1-1: General Rules and Rules for Buildings*; European Committee for Standardization (CEN): Brussels, Belgium, 2003.
13. EN 1998-1: 2004. *Eurocode 8: Design of Structures for Earthquake Resistance, Part 1: General Rules, Seismic Actions and Rules for Buildings*; CEN: Brussels, Belgium, 2003.
14. Kleinhenz, M.; Winter, S.; Dietsch, P. Eurocode 5—A halftime summary of the revision process. In Proceedings of the WCTE, Vienna, Austria, 22–25 August 2016.
15. Follesa, M.; Fragiacomo, M.; Casagrande, D.; Tomasi, R.; Piazza, M.; Vassallo, D.; Canetti, D.; Rossi, S. The New Provisions for the Seismic Design of Timber Buildings in Europe. *Eng. Struct.* **2018**, *168*, 736–747. [CrossRef]
16. Ceccotti, A. New technologies for construction of medium-rise buildings in seismic regions: The XLAM case. *Struct. Eng. Int.* **2008**, *18*, 156–165. [CrossRef]
17. Ceccotti, A.; Sandhaas, C.; Okabe, M.; Yasumura, M.; Minowa, C.; Kawai, N. SOFIE project—3D shaking table test on a seven-storey full-scale cross-laminated timber building. *Earth Eng. Struct. Dyn.* **2013**, *42*, 2003–2021. [CrossRef]
18. Van de Lindt, J.W.; Furley, J.; Amini, M.O.; Pei, S.; Tamagnone, G.; Barbosa, A.R.; Rammer, D.; Line, P.; Fragiacomo, M.; Popovski, M. Experimental seismic behaviour of a two-story CLT platform building. *Eng. Struct.* **2019**, *183*, 408–422. [CrossRef]
19. Dong, W.; Li, M.; Ottenhaus, L.; Lim, H. Ductility and overstrength of nailed CLT hold-down connections. *Eng. Struct.* **2020**, *215*, 110667. [CrossRef]
20. Fragiacomo, M.; Dujic, B.; Sustersic, I. Elastic and ductile design of multi-storey crosslam massive wooden buildings under seismic actions. *Eng. Struct.* **2011**, *33*, 3043–3053. [CrossRef]
21. Gavric, I.; Fragiacomo, M.; Ceccotti, A. Cyclic behavior of typical metal connectors for cross-laminated (CLT) structures. *Mater. Struct.* **2015**, *48*, 1841–1857. [CrossRef]
22. Pozza, L.; Saetta, A.; Savoia, M.; Talledo, D. Angle bracket connections for CLT structures: Experimental characterization and numerical modeling, Construct. *Build. Mater.* **2018**, *191*, 95–113. [CrossRef]
23. Izzi, M.; Casagrande, D.; Bezzi, S.; Pasca, D.; Follesa, M.; Tomasi, R. Seismic behavior of Cross-Laminated Timber structures: A state-of-the-art review. *Eng. Struct.* **2018**, *170*, 45–52. [CrossRef]

24. Trutalli, D.; Marchi, L.; Scotta, R.; Pozza, L. Capacity design of traditional and innovative ductile connections for earthquake-resistant CLT structures. *Bull. Earth Eng.* **2019**, *17*, 2115–2136. [CrossRef]
25. Latour, M.; Rizzano, G. Cyclic behavior and modeling of a dissipative connector for Cross-Laminated Timber panel buildings. *J. Earth Eng.* **2015**, *19*, 137–171. [CrossRef]
26. Scotta, R.; Marchi, L.; Trutalli, D.; Pozza, L. A dissipative connector for CLT buildings: Concept, design and testing. *Materials* **2016**, *9*, 139. [CrossRef] [PubMed]
27. UNI EN 14081-1:2016. *Timber Structures-Strength Graded Structural Timber with Rectangular Cross Section-Part 1: General Requirements*; CEN: Brussels, Belgium, 2001.
28. ETA-14/0349. *European Technical Assessment, Issued in Accordance with Regulation EU N° 305/2011, 2019*; European Commission: Brussels, Belgium, 2019.
29. DIN EN 1995-1-1/NA. *National Annex-Nationally Determined Parameters-Eurocode 5: Design of Timber Structures-Part 1-1: General-Common Rules and Rules for Buildings*; Deutsche Institute fur Normung (DIN): Berlin, Germany, 2013.
30. ÖNORM B 1995-1-1. *Eurocode 5: Design of Timber Structures–Part 1-1: General Common Rules and Rules for Buildings–National Specification for the Implementation of ONORM EN 1995-1-1, National Comments and National Supplements*; Austrian Standards Institute (ASI): Vienna, Austria, 2015.
31. Italian Technical Document CNR DT 2016/2018-R1. *Instruction for Design, Execution and Control and Timber Construction*; National Council of Research: Rome, Italy, 2018.
32. International Building Code (IBC). *Produced by International Code Council*; International Building Code (IBC): Washington, DC, USA, 2015.
33. American Wood Council. *National Design Specification (NDS) for Wood Construction*; American Wood Council: Leesburg, VI, USA, 2015.
34. Japan CLT Association. *Manual of CLT Construction Design for Practitioners*; Edition Detail: Osaka, Japan, 2018.
35. Australian Cross Laminated Timber Panel Structural Guide. In *Australian Design Guide*; Australian Edition: New South Wales, Australia, 2017.
36. Wallner-Novak, M.; Koppelhuber, J.; Pock, K. *Cross-Laminated Timber Structural Design-Basic Design and Engineering Principles According to Eurocode*; proHolz Austria: Vienna, Austria, 2013.
37. Shahnewaz, M.; Popovski, M.; Tannert, T. Deflection of cross-laminated timber shear walls for platform-type. *Eng. Struct.* **2020**, *221*, 111091. [CrossRef]
38. Chen, Z.; Popovski, M. Mechanics-based analytical models for balloon-type cross-laminated timber (CLT) shear walls under lateral loads. *Eng. Struct.* **2020**, *208*, 109916. [CrossRef]
39. Johansen, K.W. Theory of timber connections. *Int. Ass Bridge Struct. Eng.* **1949**, *9*, 249–262.
40. Flatscher, G.; Bratulic, K.; Schickhofer, G. Experimental tests on cross-laminated timber joints and walls. *Proc. ICE Struct. Build.* **2015**, *168*, 868–877. [CrossRef]
41. Popovski, M.; Schneider, J.; Schweinsteiger, M. Lateral load resistance of cross-laminated wood panels. In Proceedings of the WCTE, Riva del Garda, Italy, 20–24 June 2010.
42. Tomasi, R.; Smith, I. experimental characterization of monotonic and cyclic loading responses of CLT panel-to-foundation angle-brackets connections. *J. Mater. Civ. Eng.* **2015**, *27*, 04014189. [CrossRef]
43. Latour, M.; Rizzano, G. Seismic behavior of cross-laminated timber equipped with traditional and innovative connectors. *Arch. Civ. Mech. Eng.* **2017**, *17*, 382–399. [CrossRef]
44. Pozza, L.; Ferracuti, B.; Massari, M.; Savoia, M. Axial-shear interaction on CLT hold-down connections-Experimental investigation. *Eng. Struct.* **2018**, *160*, 95–110. [CrossRef]
45. Liu, J.; Lam, F. Experimental test of Cross-Laminated Timber connections under bidirectional loading. In Proceedings of the WCTE, Vienna, Austria, 22–25 August 2016.
46. Franco, L.; Pozza, L.; Saetta, A.; Savoia, M.; Talledo, D. Strategies for structural modeling of CLT panels under cyclic loading conditions. *Eng. Struct.* **2019**, *198*, 109476. [CrossRef]
47. Gavric, I.; Fragicomo, M.; Ceccotti, A. Cyclic behavior of CLT wall systems: Experimental tests and analytical prediction models. *J. Struct. Eng.* **2015**, *141*, 04015034. [CrossRef]
48. Blass, J.H.; Fellmoser, P. Design of solid wood panels with cross layers. In Proceedings of the WCTE 2004, Lahti, Finland, 14–17 June 2004; pp. 543–548.
49. Sandoli, A.; Moroder, D.; Pampanin, S.; Calderoni, B. Simplified analytical models for coupled CLT walls. In Proceedings of the WCTE, Vienna, Austria, 22–25 August 2016.
50. Sandoli, A.; D'Ambra, C.; Ceraldi, C.; Calderoni, B.; Prota, A. Role of perpendicular to grain compression properties on the seismic behavior of CLT walls. *J. Build. Eng.* **2021**, *34*, 101889. [CrossRef]
51. Blass, H.J.; Fellmoser, P. Influence of rolling shear modulus on strength and stiness of structural bonded timber elements. In Proceedings of the CIB-W18 Meeting, Edinburg, TX, USA, August 2004.
52. Sandoli, A.; Calderoni, B. The rolling shear influence on the out-of-plane behavior of CLT panels: A comparative analysis. *Buildings* **2020**, *10*, 42. [CrossRef]
53. Corpataux, L.; Okuda, S.; Kua, W.H. Panel and plate properties of Cross-laminated timber (CLT) with tropical fast-growing timber species in compliance with Eurocode 5. *Constr. Build. Mater.* **2020**, *261*, 119672. [CrossRef]

54. Lim, H.; Tripathi, S.; Li, M. Rolling shear modulus and strength of cross-laminated timber treated in micronized copper azole type C (MCA-C). *Constr. Build. Mater.* **2020**, *259*, 120419. [CrossRef]

55. Minghao, L. Evaluating rolling shear strength properties of cross-laminated timber by short-span bending tests and modified planar shear tests. *J. Wood Sci.* **2017**, *63*, 331–337.

56. Zhou, Q.; Gong, M.; Chui, Y.H.; Mohammad, M. Measurement of rolling shear modulus and strength of cross laminated timber fabricated with black spruce. *Constr. Build. Mater.* **2017**, *64*, 379–386. [CrossRef]

57. Saavedra Flores, E.J.; Saavedra, K.; Hinojosa, H.; Chandra, Y.; Das, R. Multi-scale modeling of rolling shear failure in cross-laminated timber structures by homogenization and cohesive zone models. *Int. J. Solid Struct.* **2016**, *81*, 219–232. [CrossRef]

58. Sturzenbecher, R.; Hofstetter, K.; Eberhardsteiner, J. Structural design of Cross Laminated Timber (CLT) by advanced plate theories. *Comp. Sci. Technol.* **2010**, *70*, 1368–1379. [CrossRef]

59. Kreuzinger, H. Mechanically jointed beams and columns. *Timber Eng.* **1995**, *1*, B11.

60. Casagrande, D.; Rossi, S.; Sartori, T.; Tomasi, R. Proposal of an analytical procedure and a simplified numerical model for elastic response of single-storey timber shearwalls. *J. Constr. Build. Mater.* **2016**, *102*, 1101–1112. [CrossRef]

61. Gavric, I.; Popovski, M. Design models for CLT shear walls and assemblies based on connection properties. *Int. Netw. Timber Eng. Res.* **2014**, *15*, 267–280.

62. Pei, S.; Lindt, J.V.D.; Popovski, M. Approximate R-factor for cross-laminated timber walls in multi-story buildings. *J. Arch. Eng.* **2012**, *19*, 245–255. [CrossRef]

63. Tomasi, R. *CLT Course at FPS COST Action FP1004–Enhance Mechanical Properties of Timber, Engineered Wood Products and Timber Structures*; CLT Training School, University of Trento: Trento, Italy, 2014.

64. Tamagnone, G.; Rinaldin, G.; Fragiacomo, M. A novel method for non-linear design of CLT wall systems. *Eng. Struct.* **2018**, *167*, 760–771. [CrossRef]

65. Franco, L.; Pozza, L.; Saetta, A.; Talledo, D. Enhanced N-V interaction domains for the design of CLT shear wall based on coupled interaction models. *Eng. Struct.* **2021**, *231*, 111607. [CrossRef]

66. Pozza, L.; Trutalli, D. An analytical formulation of the q-factor for mid-rise CLT buildings based on parametric numerical analyses. *Bull. Earth Eng.* **2015**, *13*, 3449–3469. [CrossRef]

67. Tran, K.D.; Jeong, Y.G. Effects of wood species, connection system, and wall-support interface type on cyclic behaviours of cross-laminated timber (CLT) walls under lateral loads. *Constr. Build. Mater.* **2021**, *280*, 122450. [CrossRef]

68. Yasumura, M.; Kobayashi, K.; Okabe, M.; Miyake, T.; Matsumoto, K. Full-scale tests and numerical analysis of low-rise CLT structures under lateral loading. *J. Struct. Eng.* **2016**, *142*, E4015007. [CrossRef]

69. Izzi, M.; Polastri, A.; Fragiacomo, M. Investigating the hysteretic behavior of Cross- Laminated Timber wall systems due to connections. *J. Struct. Eng.* **2018**, *144*, 04018035. [CrossRef]

70. Mestar, M.; Doudak, G.; Polastri, A.; Casagrande, D. investigating the kinematic modes of CLT shear-walls with openings. *Eng. Struct.* **2021**, *228*, 111475. [CrossRef]

71. Moroder, D.; Sarti, F.; Pampanin, S.; Smith, T.; Buchanan, A.H. Higher mode effects in multi-storey timber buildings with varying diaphgrams flexibility. In Proceedings of the NZSEE Conferernce, Christchurch, New Zealand, 1–3 April 2016.

72. Moroder, D.; Smith, T.; Armstrong, J.; Young, B.; Buchanan, A.H. Challenges and solutions in the design of a 10 storey CLT building. *N. Zealand Timber Design* **2017**, *25*, 4.

73. Flaig, M.; Blass, H.J. Shear strength and shear stiffness of CLT-beams loaded in plane. In Proceedings of the CIB-W18 Meeting 46, Vancouver, BC, Canada, 26–29 August 2013. paper 46-12-3.

74. Applied Technology Council. *ATC-7 Guidelines for the Design of Horizontal Wood Diaphgrams*; Applied Technology Council: Berkeley, CA, USA, 1981.

75. Vassallo, D.; Follesa, M.; Fragiacomo, M. Seismic design of a six-storey CLT building in Italy. *Eng. Struct.* **2018**, *175*, 322–338. [CrossRef]

76. Christovasilis, I.P.; Riparbelli, L.; Rinaldin, G.; Tamagnone, G. Methods for practice-oriented linear analysis in seismic design of Cross Laminated Timber buildings, Soil Dynam. *Earthq. Eng.* **2020**, *128*, 105869.

77. Rinaldin, G.; Amadio, A.; Fragiacomo, M. A component approach for the hysteretic behavior of connections in cross-laminated wooden structures. *Earth Eng. Struct. Dyn.* **2013**, *42*, 2023–2042. [CrossRef]

78. Masroor, M.; Doudak, G.; Casagrande, D. The effect of bi-axial behavior of mechanical anchors on the lateral response of multi-panel CLT sharwalls. *Eng. Struct.* **2020**, *224*, 111202. [CrossRef]

79. Schneider, J.; Karacabeyli, E.; Popovski, M.; Steimer, S.F.; Tesfamariam, S. Damage assessment of connections used in Cross-Laminated Timber subjected to cyclic loads. *J. Perform Facil.* **2014**, *28*, A4014008.

80. Piestley, N.J.M.; Tao, R.J. Seismic Response of Precast Prestressed Concrete Frames with Partially Debonded Tendons. *PCI J.* **1993**, *38*, 58–69. [CrossRef]

81. Palermo, A.; Pampanin, S.; Buchanan, A.; Newcombe, M. Seismic design of multi-storey buildings using laminated veneer lumber (LVL). In Proceedings of the New Zealand Earthquake Engineering conference NZEES, Christchurch, New Zealand, 11–13 March 2005.

82. Sarti, F.; Palermo, A.; Pampanin, S. Quasi static cyclic testing of two-thirds scale un-bonded posttensioned rocking dissipative timber walls. *J. Struct. Eng.* **2016**, *142*, E40115005.

applied sciences

 MDPI

Article

Interoperability between Building Information Modelling (BIM) and Building Energy Model (BEM)

Gabriela Bastos Porsani, Kattalin Del Valle de Lersundi, Ana Sánchez-Ostiz Gutiérrez and Carlos Fernández Bandera *

School of Architecture, University of Navarra, 31009 Pamplona, Spain; gbastospors@alumni.unav.es (G.B.P.); kdelvalle@unav.es (K.D.V.d.L.); aostiz@unav.es (A.S.-O.G.)
* Correspondence: cfbandera@unav.es; Tel.: +34-948-425-600 (ext. 803189)

Abstract: Building information modelling (BIM) is the first step towards the implementation of the industrial revolution 4.0, in which virtual reality and digital twins are key elements. At present, buildings are responsible for 40% of the energy consumption in Europe and, so, there is a growing interest in reducing their energy use. In this context, proper interoperability between BIM and building energy model (BEM) is paramount for integrating the digital world into the construction sector and, therefore, increasing competitiveness by saving costs. This paper evaluates whether there is an automated or semi-automated BIM to BEM workflow that could improve the building design process. For this purpose, a residential building and a warehouse are constructed using the same BIM authoring tool (Revit), where two open schemas were used: green building extensible markup language (gbXML) and industry foundation classes (IFC). These transfer files were imported into software compatible with the EnergyPlus engine—Design Builder, Open Studio, and CYPETHERM HE—in which simulations were performed. Our results showed that the energy models were built up to 7.50% smaller than in the BIM and with missing elements in their thermal envelope. Nevertheless, the materials were properly transferred to gbXML and IFC formats. Moreover, the simulation results revealed a huge difference in values between the models generated by the open schemas, in the range of 6 to 900 times. Overall, we conclude that there exists a semi-automated workflow from BIM to BEM which does not work well for big and complex buildings, as they present major problems when creating the energy model. Furthermore, most of the issues encountered in BEM were errors in the transfer of BIM data to gbXML and IFC files. Therefore, we emphasise the need to improve compatibility between BIM and model exchange formats by their developers, in order to promote BIM–BEM interoperability.

Keywords: building information modelling (BIM); building energy model (BEM); green building extensible markup language (gbXML); industry foundation classes (IFC); interoperability; digital twin; sustainable construction; intelligent buildings assessment; sustainability performance; simulation tools for building

Citation: Bastos Porsani, G.; Del Valle de Lersundi, K.; Sánchez-Ostiz Gutiérrez, A.; Fernández Bandera, C. Interoperability between Building Information Modelling (BIM) and Building Energy Model (BEM). *Appl. Sci.* **2021**, *11*, 2167. https://doi.org/10.3390/app11052167

Academic Editor: Elmira Jamei

Received: 31 January 2021
Accepted: 23 February 2021
Published: 1 March 2021

1. Introduction

The challenge of designing a high-performance building demands better data exchange between the building information modelling (BIM) and building energy model (BEM). The building sector is responsible for 40% of the energy consumption in Europe, in which the building envelope is the main constructive element, in terms of its impact on the building's energy consumption [1]. The European Union has submitted a set of directives to eliminate inefficient buildings, through the Energy Performance of Buildings Directive (EPBD) [2,3], which is the leading legislative and policy instrument. This directive focuses on the building sector (i.e., new and existing constructions) and determines that the existing ones must be nearly zero energy buildings (NZEBs) by 2050 [4]. The NZEB standard is valuable for the environment, as well as for the end-users of the building, as energy savings also reduce costs [5]. In this framework, there has been a growing interest in energy modelling studies, in order to understand how to optimise energy consumption in buildings [6].

In this context, on the 28 October 2020, the Sustainable Places 2020 virtual conference took place, where nine EU-funded research projects showed their work and exposed some innovative solutions advanced in their projects. All of them highlighted the challenges that the industry continues to face under the digital transformation [7]. Recently, the construction sector has been identified as the poorest in Europe, in terms of productivity, as there exist many difficulties in this sector to integrate digital innovations [8].

Over the last two decades, BIM has been used in the construction sector, with the focus to improve collaboration between different stakeholders [9], facilitate the fast creation of different designs, and to decrease its inefficiencies on-site and among the project disciplines [10]. It has also become the core of the cyber-physical system and a great tool to improve the energy efficiency concept in building design, once it is able to improve the quality and performance of the construction life-cycle under the industry 4.0 paradigm [11]. At present, BIM is crucial for bridging the gap between digitalisation and the construction field. The digital information management tools offered by BIM make it possible to optimise and better control constructive processes, from the design stage to the management of the building's life-cycle and maintenance [12].

BEMs are the basis of energy performance certificates (EPCs) and assessments. Both EPCs and assessments should be user-friendly, profitable, and more reliable, in order to give investors confidence in the energy efficiency sector. Then, the certification processes and BEMs should achieve these requirements [13]. In addition, BEMs can be organized into three categories: Physics-based models (white-box) [14], pure data-driven or statistical models (black-box) [15], and hybrid models (grey-box) [16]. In this paper, white-box models were considered, as they enable the definition of all the elements that influence the building's energy behaviour (e.g., the weather, internal loads, construction) [17,18] and, therefore, have a clear connection with BIM components. Although heating ventilation air conditioning (HVAC) systems are also important to the energy analysis, they were not considered in this study.

Furthermore, interoperability between BIM and BEM could be a step forward in reducing costs and saving model re-creation time during the design phase. BIM has the potential to streamline the production of energy models, by storing important data such as building geometry, construction typology, and thermal properties [19], as well as enabling the iteration of efficient designs in a shared environment [20]. In addition, BEMs are key elements in understanding how to reduce energy consumption and how to improve it [21]. Furthermore, BEM tools and measures can carry out energy simulations, analyse energy needs, and improve the project design [22]. Due to this, at present, there is a significant interest in simplifying the creation of a BEM from the BIM authoring tool, by sharing and exporting the data from the architectural model [23,24].

However, BIM–BEM interoperability is one of the existing digital gaps at the design stage. Although BIM is considered a multidisciplinary tool, interoperability issues still prevent many BIM applications in relevant industries [25]. The lack of compatibility between the BIM model and the energy simulation applications, the repetitive manual operations required to create a BEM, and the non-standardised and subjective process [26] usually result in data loss and misinterpretation, especially when this was been taken into account in the first phase of the design step [27,28]. Sanhudo et al. also concluded that the main problem in transferring data between BIM and BEM is the information loss, which can lead to a rework consisting of re-entering the BIM stored information into the energy model [29]. Furthermore, Kamal and Memari were in accordance with Sanhudo et al., defending that BIM contributes to the easy handling of data, which can leads to automation in building energy modelling and simulation, as well as providing up-to-date energy models with real-time information. They also emphasised the need for clear standards and solutions for the BIM to BEM interoperability process [30].

The BIM–BEM process consists of three parts—BIM tool, model schema exchange format, and BEM software—interoperability issues may appear in any or all of these; not only in the energy simulation application. This means the problems that are encountered are not exclusively due to the limited ability of BEM tools to translate input data from BIM [31]. In order to transfer the BIM information to a building energy analysis software, many file formats exist, such as hypertext markup language (HTML), extensible hypertext markup language (XHTML), building construction extensible markup language (bcXML), industry foundation classes extensible markup language (IFCXML), industry foundation classes (IFC), and green building extensible markup language (gbXML), the last two being the main open BIM standards [32]. Open standards enable and facilitate the integration and collaboration of different professionals from all over the world, by providing a common basis around which research, development, and deployment activities can be organised, as has been done in the IBPSA Project 1 [33].

These open standards are commonly used in the BIM–BEM procedure, and there is an ongoing discussion about which standard is the best for the data exchange process. On the one hand, IFC reproduces a complete building project and, therefore, generates a complex data schema and larger size data files. On the other hand, gbXML provides a more flexible and direct approach to energy analysis [34]. Both of them transfer the material properties, data for HVAC systems, and thermal zones; however, only gbXML provides location data [30]. However, Gao et al. drew different conclusions about gbXML and IFC data transfer. They invented a classification process called BIM-based-BEM, which divides the information transfer from BIM into six categories: Geometry (step 1), material (step 2), space type (step 3), thermal zone (step 4), space load (step 5), and HVAC (step 6). After analysing the information transfer between distinctive softwares, they verified that the IFC format is step 1 and gbXML is step 3. In addition, they concluded that the information transfer is not user-friendly, as there are many steps between BIM and BEM [35,36]. In summary, BIM–BEM interoperability has not yet become as widespread as expected, due to the information loss, lack of a standard, software not being as interoperable as claimed by their suppliers, and the data standards not having been clearly determined [37].

Due to the need to improve BIM–BEM interoperability in the design phase, this work aims to verify and evaluate whether there is an automated or semi-automated BIM to BEM workflow, comparing the importing of gbXML and IFC in three BEM software: Design Builder, Open Studio, and CYPETHERM HE. These software were chosen as they are all based on EnergyPlus [38], which is a notable and reliable energy simulation engine. EnergyPlus does not have a user-friendly interface; therefore, these three different software were chosen to run the engine. Firstly, Design Builder [39] is the most global tool, used by most professionals in the simulation sector. Secondly, Open Studio [40] was chosen because it is an open-source software, although it does not have the ability to model from scratch; instead relying on SketchUp to do so which, by contrast, is no longer a free software. Both Design Builder and Open Studio were created for energy simulations and they are the most popular tools for BEM creation. Finally, CYPETHERM HE [41] is a tool that has wide international outreach and is part of the complete CYPE package, made up of other software related to architecture, engineering, and construction. Furthermore, this software is BIM-oriented and should have a good interoperability process.

Furthermore, we intend to lead to an improvement in the building design stage through this research, in terms of speeding it up, obtaining more precise energy data, and encouraging professionals in the construction field to gain confidence in the process and helping them to be aware of the limitations. This work could as well be a guide for BIM/BEM software developers, in order to find new paths to overcome the main problems that the process is now facing.

For this purpose, two case studies at different scales were chosen: A single-family house and a warehouse. In both cases, we started from a BIM model and explored different paths to reach the BEM. The difficulties and the problems detected in the thermal envelope in the interoperability process were investigated.

The rest of this paper is organized as follows: Section 2 explains the method, which consists of the BIM exporting formats and the BEM importing software. Section 3 analyses and discusses the results. Section 4 presents the conclusion and research directions for future works.

2. Method

The aim of this research was to verify whether there is an automated workflow between building information modelling (BIM) and building energy model (BEM) software, and how this interoperability works. The idea was to experiment with three different flows from BIM to BEM, which had the same BIM authoring tool in common and the same calculation engine in the BEM software: Energy Plus. For this study, 3D models were made in Autodesk Revit 2020 and the three BEM programs that were chosen were Design Builder, Open Studio, and CYPETHERM HE. Figure 1 represents the paper's structure and process.

The paper was divided into two main steps: The first one is about the interoperability from BIM to BEM for two case studies, a residential building and an industrial warehouse. For both projects, the files were exported from Revit in two formats: green building extensible markup language (gbXML) and industry foundation classes (IFC). To export the gbXML file, we needed to create an analytical model in Revit, which was not necessary for the IFC format. However, to import the IFC file into CYPETHERM HE, another software was required (Open BIM Analytical Model), in order to generate an analytical model based on the IFC format without errors. At the end of this step, a comparison of the gbXML and IFC import results was made to identify the interoperability strengths and weaknesses. In addition, a qualitative evaluation of this first step was carried out, in order to verify which of the three processes was more feasible, under the vision of an architect user of the software.

The last step was a comparison of the simulation results of the first step. The extracted files from BIM were imported into the BEM software: Design Builder and Open Studio were used for gbXML and CYPETHERM HE for IFC. To make the comparison more precise, we focused on the energy envelope performance through windows and opaque surfaces. In this way, none of the data about people, equipment, lights, infiltration, or HVAC were selected and, therefore, we could explore only the parameters affected by the interoperability construction process in more detail.

As a result, we intended to discern whether there exists an automated or semi-automated workflow between BIM and BEM.

2.1. BIM Exporting Formats

At present, many BIM authoring tools support the gbXML and IFC data formats. In this paper, we chose Autodesk Revit 2020 as the BIM engine, in which only the geometry and envelopes of the buildings were modelled. As mentioned previously, HVAC was not considered in the BEM.

The first case study is a residential building of 143.60 m², with two floors (Figure 2) and 13 thermal zones. It has a simple and straight-line geometry, with vertical circulation per staircase. Its Revit file had 4.20 megabytes of information.

On the other hand, the second case study was a much bigger and more complex building (Figure 2). It is an industrial warehouse of 9.677 m², with three floors and a basement, in which 21 thermal zones were defined. Although it has a simple geometry, the roof is curved, as well as some of the walls. In comparison with the residential case, this one had 75.15 megabytes of information.

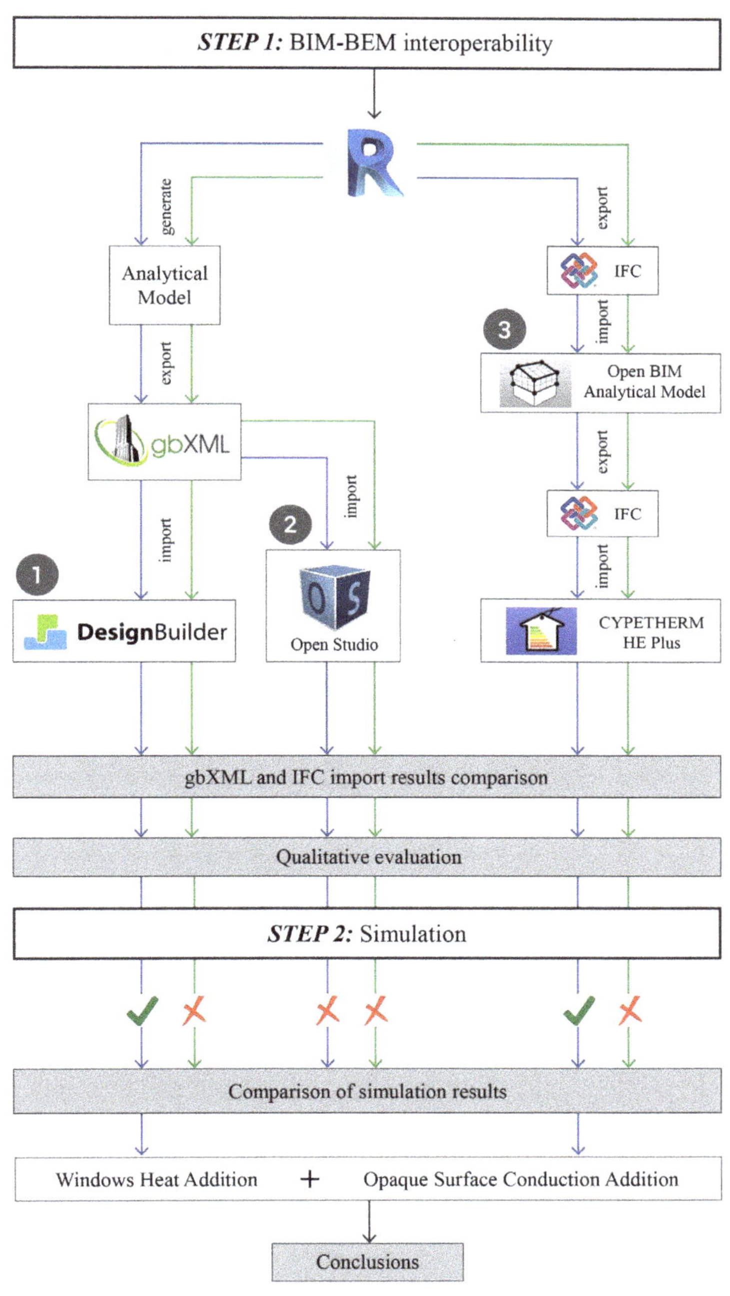

Figure 1. Paper process diagram. Numbers to follow the diagram and import processes.

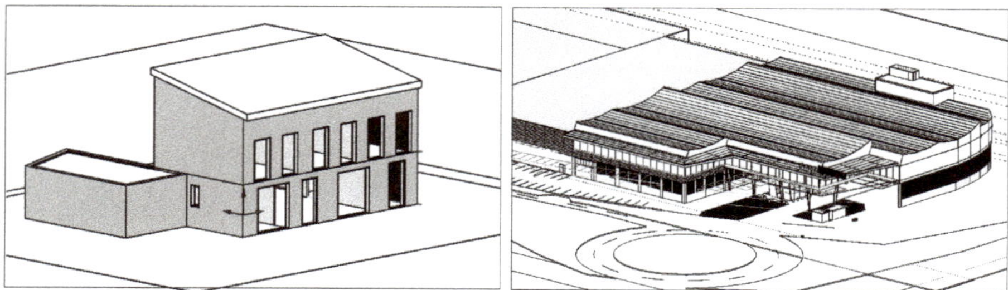

Figure 2. Building information modelling (BIM) models of the residential and warehouse buildings.

2.1.1. Green Building XML (gbXML)

The Green Building XML (gbXML) schema, developed by Green Building Studio Inc., is focused on environmental data. It was originally developed to enable interoperability between the building design in CAD softwares and energy analysis tools [34]. However, gbXML also attends to exchange data among diverse simulation instruments. In its format, the geometric and weather data, user profiles, and energy information are exchanged. Nevertheless, some parameters about building HVAC components are discarded [42].

At the moment of exporting the gbXML file from Revit, some decisions must be made. First, it is necessary to create an analytical model, and to check, in the Revit export gbXML interface, if the spaces and the geometry are defined precisely. If there is any error with both characteristics, it would not be able to conclude the gbXML export. Secondly, in order to achieve good interoperability, some parameters must also be activated in Revit export gbXML, such as rooms and detailed elements; otherwise, in the gbXML, there will be a lack of information needed for the creation of the BEM, in reference to the geometry and the envelope materials.

It is possible to see, from Figure 3, that all the thermal zones of the residential building were generated correctly in the gbXML file. Contrarily, the warehouse study presented a particularity in its analytical model and in the gbXML export. Its analytical spaces' 3D view did not show all the thermal zones that were created. However, at the point of the gbXML export, it can be seen that the zones were generated properly.

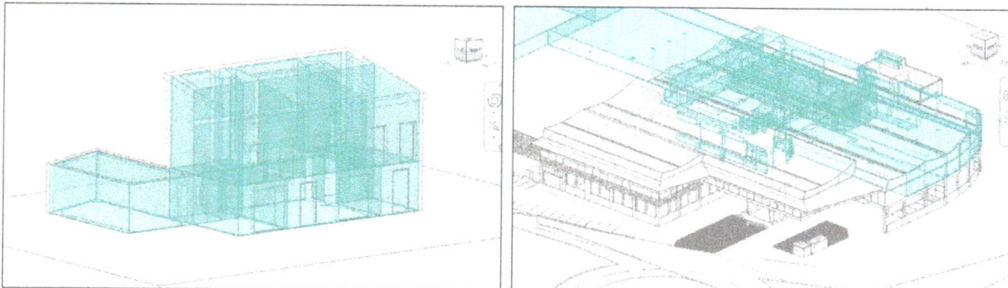

Figure 3. Analytical model of the residential and warehouse buildings.

2.1.2. Industry Foundation Classes (IFC)

The industry foundation classes (IFC) were developed by international association interoperability (IAI), but are now administrated by the buildingSMART alliance. They are the only 3D object-oriented open standards that use BIM. In contrast to gbXML, the aim of the IFC format is to provide a single basis for the exchange of information in the field of construction and facility management [43]. The IFC schema is more extensive and complex than the gbXML format, and supports, in addition to a good geometric representation, semantic data enrichment [44].

Although there is an automated IFC export from Revit using the BIM collaboration plug-in developed by CYPE, the IFC files for both case studies were exported manually. In order to guarantee proper interoperability, in the Revit export interface, the phase where the thermal zones were created must be selected. Otherwise, the room data will not be transferred.

2.2. BEM Importing Software

As mentioned above, Design Builder and Open Studio were the software selected to import the gbXML file, while CYPETHERM HE was used to import the IFC format.

2.2.1. Design Builder

First of all, the gbXML file was imported in Design Builder, which is a software that allows for the editing of BEM geometry, spaces, and materials. In this research, as the purpose was to study the interoperability from BIM to BEM, any modification was made in the imported model. Nevertheless, at the moment of the gbXML file insert, some parameters were activated and some templates were changed, as data related to the typology template, weather file, ideal loads, and occupation schedule were not detailed in Revit and, therefore, could not be imported with the gbXML file.

In the interface, "Import BIM Model" was selected to import the thermal properties and the model as building blocks. The use of these parameters guaranteed that the materials defined in Revit would be transferred into Design Builder and that the model would be constructed with thermal zones. The activity template chosen was according to the type of the building and, in the openings tab, the layout of the external windows was defined as none. If these elements had not been defined properly, the BEM model would not work correctly and this could not be considered an interoperability error but, rather, a mistake in the configuration of the BEM file.

2.2.2. Open Studio

In order to compare the transfer of the data, the gbXML file was imported into the Open Studio software. Differently to Design Builder, Open Studio only allows for the change of the materials and the thermal zones and, if any adjustment to the geometry is needed, the SketchUp Open Studio plug-in software must be used. Similarly to Design Builder, it was also necessary to carry out some actions in Open Studio, such as changing the typology template, choosing the appropriate timetable, activating the ideal loads, and selecting the weather file.

2.2.3. CYPETHERM HE

Initially, three options for BEM software were considered to import IFC: CYPETHERM HE, Space Boundary Tool (SBT), and Open Studio. Nevertheless, only CYPETHERM HE was actually available. The SBT software, from the Lawrence Berkeley National Laboratory, is no longer updated (since 2014). On the other hand, Open Studio can import gbXML, as was previously explained, but when trying to import IFC files, an intermediate program called BIMserver was necessary. The installation process of the BIMserver required a high level of expertise and, therefore, this workflow was discarded.

CYPETHERM HE is continuously updated and its interoperability between BIM and BEM is an easy and intuitive process. Despite the fact that an intermediate software is needed to transfer the IFC data to CYPETHERM HE (Open BIM Analytical Model), the workflow is simple and does not require expert knowledge.

In the CYPETHERM HE import interface, it is possible to select a typology directory, which refers to the project materials. In this case, as it was not possible to transfer the envelope materials and its thermal properties from Revit and there was no typology directory created in CYPETHERM HE, it was necessary to generate all the construction systems in this tool. Once this typology is generated, it can be saved and be used in any other project.

Before importing the IFC file into the BEM tool, the Open BIM Analytical Model software was used to create an analytical model and verify the extracted data, as it refers to its dimensions, geometry, materials, and any inconsistency between the building geometry elements. As a result, the IFC data of the residential case was extracted correctly from the BIM authoring tool. All the 13 thermal zones and the envelope materials were transferred. Furthermore, the model was constructed properly and its dimensions were the same as in the BIM (see Figure 4).

On the other hand, the industrial warehouse model in the Open BIM Analytical Model software seemed to be constructed properly, but only 19 thermal zones were imported. It was found out that two thermal zones of the first floor were missing in the IFC data. No reason was found, as to why these zones were not extracted, as the same procedure was followed within all the thermal zones.

In order to cast light on the problem, the IFC of the warehouse was imported into the Solibri Anywhere software and the same issues appeared as in the Open BIM Analytical Model. The reason for this seemed to be connected with the generation of the model from the IFC as a consequence of a wrong transfer of the BIM data to the IFC format and, therefore, was not dependent on the type of software used. This is clearly an interoperability problem related to the IFC format.

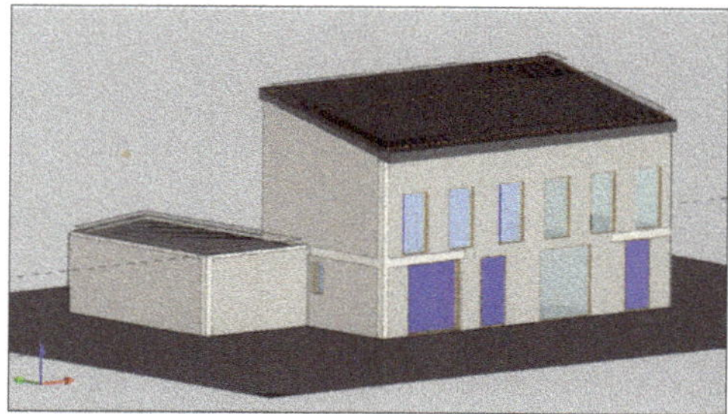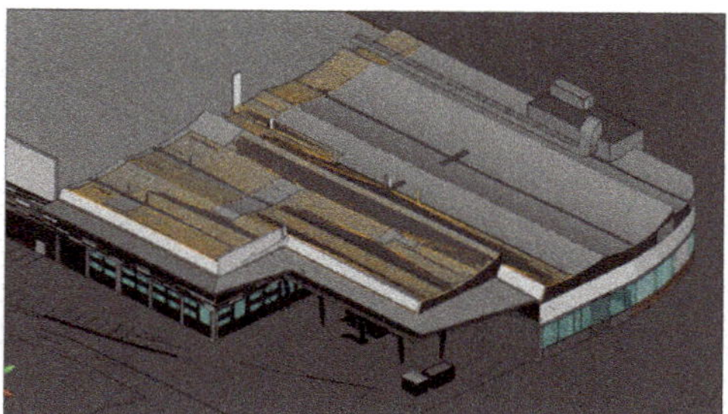

Figure 4. IFC files imported into Open BIM Analytical Model.

3. Results and Discussion

3.1. gbXML Import Results

3.1.1. Residential case in Design Builder

As it is possible to see in Figure 5, all the thermal zones established in the Revit file of the residential building were correctly imported into Design Builder. Furthermore, the library of materials was successfully transferred and a new template called "gbXML" was generated.

However, although the model was perfectly constructed in Revit, it had some problems in Design Builder. First, the model dimensions in the BEM software were not the same as in BIM. In Design Builder, the building was 1.70% smaller than in Revit and, consequently, the volumes of the thermal zones were also smaller. Another problem was found concerning the model surfaces. In the north facade, there were errors in recognizing the walls and the roof and the first floor was not constructed accurately. In addition, some of the walls were imported without the construction assignment.

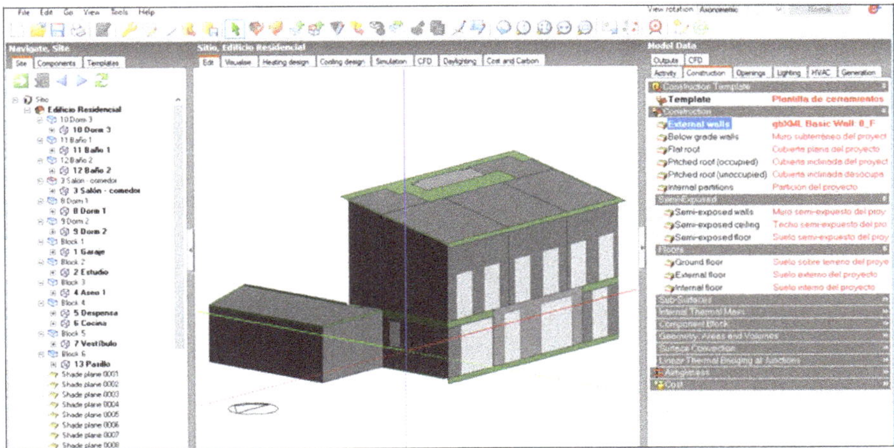

Figure 5. Residential green building extensible markup language (gbXML) file imported in Design Builder.

3.1.2. Warehouse Case in Design Builder

On the other hand, the industrial warehouse gbXML file showed many more problems than the first case study. Some walls and roof construction systems were created as shadow elements, presented apertures in their surfaces, and did not assemble well with each other. Furthermore, the roof was divided into many small parts, some of which were recognized as walls or as windows, while some of them were missing (see Figure 6). Moreover, some walls were generated automatically, although they did not exist in the BIM tool.

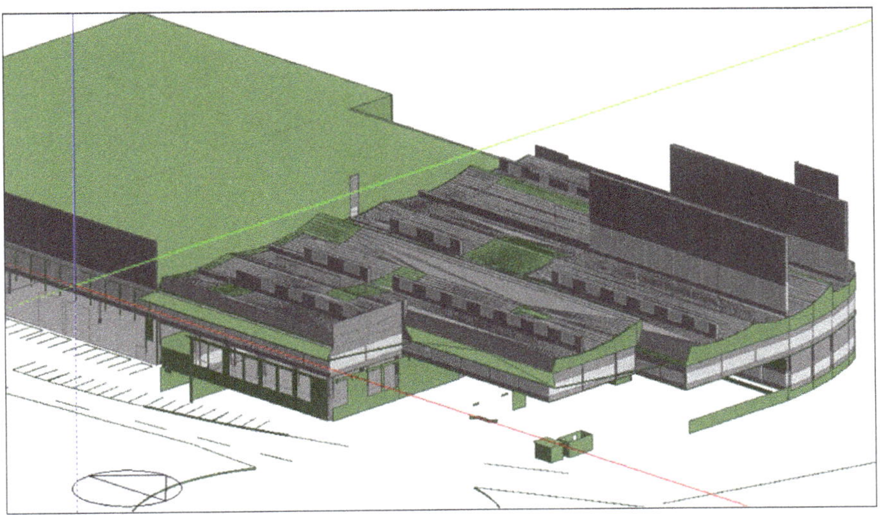

Figure 6. Warehouse gbXML file imported in Design Builder.

Table 1 summarizes the errors encountered in the gbXML file import into Design Builder. In summary, the issues found were very similar in both models. However, it was noticed that, when the file was bigger and had more data, more problems appeared during model creation in BEM, mainly with respect to the construction of the thermal envelope (walls, floors, and roof). Moreover, both models showed different dimensions from Revit, which could not be corrected at the importing step. The only way to guarantee that the model has exactly the same external dimensions as in the BIM, would be to not import it as "building blocks". Otherwise, all of the model would be created as a building component, without the generation of thermal zones. Therefore, it is better to accept a percentage of dimension error than to have to build all the thermal zones, which were already made in the BIM authoring tool.

Table 1. Comparative table between the gbXML files imported in Design Builder.

		Design Builder	
		gbXML from Revit	
		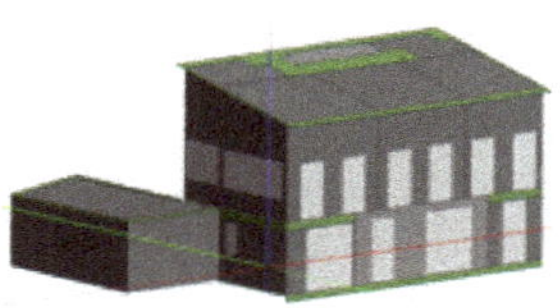	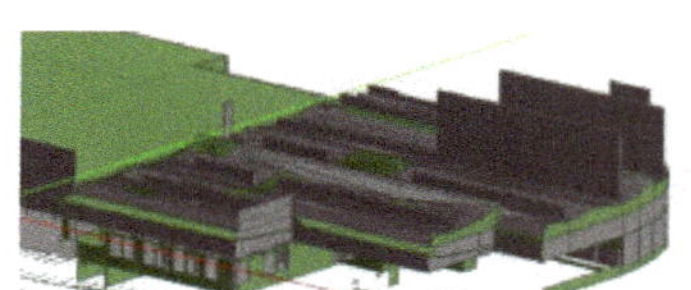
	The model	**Residential case**	**Warehouse case**
Dimensions	1. External dimensions.	Some walls were 4% smaller and others were 2% bigger than BIM	Walls 0.56% smaller than BIM.
	1. Solution.	Modify the model in Design Builder.	Modify the model in Design Builder.
	2. Thermal zones area.	Some areas were 4% smaller and other up to 10% bigger than in the BIM.	Correct.
	2. Solution.	Modify the model in Design Builder.	-
	3. Thermal zones volume.	The volumes are at least 6% bigger than in the BIM.	Correct.
	3. Solution.	Modify the model in Design Builder.	-
Geometry	4. Zones recognition.	Correct. Total of 13 zones.	Correct. Total of 21 zones.
	4. Solution.	-	-
	5. Building surfaces.	North walls and parts of the roof improperly constructed and presented some gaps. First floor is missing. Parts of the thermal envelope were recognized as shadow elements.	North walls and parts of the roof improperly constructed and presented some gaps. Ground floor is missing. Parts of the thermal envelope were recognized as shadow elements.
	5. Solution.	Modify the model in Design Builder.	Modify the model in Design Builder.
Materials	6. Building materials.	Correct. Two folders called "Imported" were created: One for constructive systems and another for materials.	Correct. Two folders called "Imported" were created: One for constructive systems and another for materials.
	6. Solution.	-	-
	7. Material thickness.	Correct.	Correct.
	7. Solution.	-	-
	8. Material thermal properties.	Correct. The properties that do not appear in Revit were automatically filled in Design Builder.	Correct. The properties that do not appear in Revit were automatically filled in Design Builder.
	8. Solution.	-	-

3.1.3. Residential Case in Open Studio

As was mentioned above, Open Studio was also used to import the gbXMl file, and some steps were executed. First, it was verified that the imported data of the materials and the thermal zones were in line with those determined in Revit. Second, with respect to the errors, it was not possible to measure the model facades in Open Studio. Then, the SketchUp 2017 software and the plug-in Open Studio for SketchUp were used to import the Open Studio file, in order to check whether the model dimensions were the same in the BIM and BEM software. However, the building dimensions were between 1% to 6% smaller and the thermal zones were 0.85% to 8% bigger than in Revit. These problems may have occurred because of the transferred data of the wall dimensions, if they were automatically recognized from external to external walls or from the middle of the internal walls. Unfortunately, this recognition cannot be controlled, either in the export of the gbXML or during its import into the BEM software.

Finally, the geometry was not generated properly: As in Design Builder, the first floor and some parts of the wall of the north facade were recognized as shadow elements. This might have occurred because it is only necessary to model the building surfaces that enclose the thermal zones and the floor is modeled as a unique line in the BEM tool. Therefore, as the model was generated automatically, every space that was left over as a shadow element was recognized. In addition, some walls had gaps that did not appear in Revit (see Figure 7).

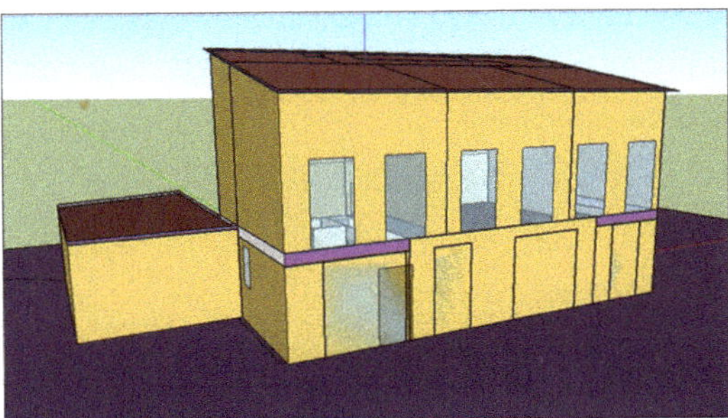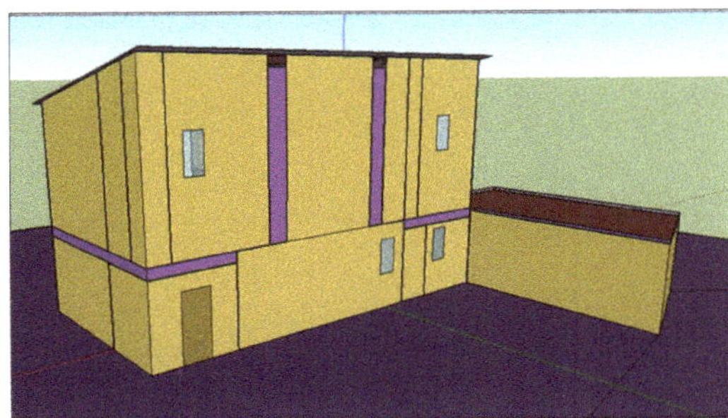

Figure 7. Residential gbXML file imported in Open Studio.

3.1.4. Warehouse Case in Open Studio

The model of the industrial warehouse in Open Studio was built with less inconsistencies than in Design Builder and all 21 thermal zones were imported. However, it also showed some errors in the construction of the walls and in its assembly with other constructive systems, which directly affected the thermal zone areas and volumes. The roof, again, was divided into many pieces, in which some parts were considered as shadow elements. Additionally, the windows were missing (see Figure 8).

As the same phenomena occurred in Design Builder, it can be noted that, in the gbXML import in Open Studio, the bigger and more complex the building, the more issues appear (mainly on the envelope). It would not be a problem if the model had the same size as the residential case, as there would not be many difficulties to be corrected. On the other hand, if the project had the same proportion or bigger than the warehouse, it would be complicated to verify all the model issues and to fix them. Finally, Table 2 compiles the errors and possible resolutions for each case.

Figure 8. Warehouse gbXML file imported in Open Studio.

Table 2. Comparative table between the gbXML files imported in Open Studio.

		Open Studio	
		gbXML from Revit	
	The model	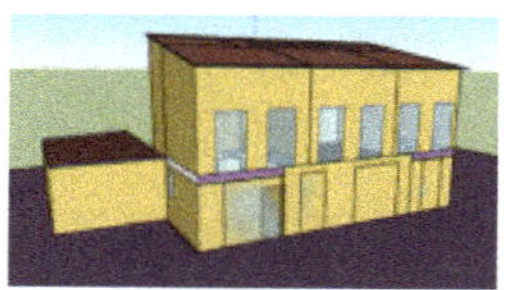 **Residential case**	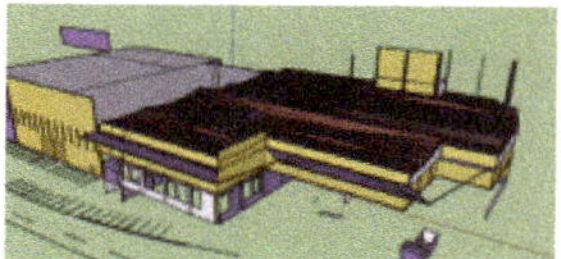 **Warehouse case**
Dimensions	1. External dimensions.	Model between 1% to 6% smaller than in the BIM.	Correct.
	1. Solution.	Modify the model in SketchUp.	-
	2. Thermal zones area	The areas are between 0.85% to 8% bigger than in the BIM.	As the thermal zones were subdivided and some parts were missing, it was not possible to measure its area.
	2. Solution.	Modify the model in SketchUp.	Modify the model in SketchUp.
	3. Thermal zones volume.	The volumes are between 4% to 12% bigger than in the BIM.	As the thermal zones were subdivided and some parts were missing, it was not possible to measure its area.
	3. Solution.	Modify the model in SketchUp.	Modify the model in SketchUp.
Geometry	4. Zones recognition.	Correct. Total of 13 zones.	Correct. Total of 21 zones.
	4. Solution.	-	-
	5. Building surfaces.	North walls and parts of the roof improperly constructed and presented some gaps. Facade of the first floor transferred as a shadow element.	Walls and parts of the roof improperly constructed, divided into many small parts and presented some gaps. Windows were missing.
	5. Solution.	-	-

Table 2. *Cont.*

		Open Studio	
Materials	6. Building materials.	Correctly imported, but some walls do not have the same construction name as in Revit.	Not all the construction systems were transferred and some walls do not have the same construction name as in Revit.
	6. Solution.	Change the name in Open Studio or in SketchUp.	Modify the model in Open Studio or in SketchUp.
	7. Material thickness. Correct.		For those materials that were imported: Correct.
	7. Solution.	-	-
	8. Material thermal properties.	In Open Studio, there are some properties that do not appear in Revit and, in this case, they were automatically filled in by Open Studio.	In Open Studio, there are some properties that do not appear in Revit and, in this case, they were automatically filled in by Open Studio.
	8. Solution.	-	-

3.2. IFC Import Results

3.2.1. Residential Case in CYPETHERM HE

Although CYPETHERM HE enables IFC import directly in the software, it has been analysed that it is better to use the IFC exported from an Open BIM Analytical Model. When the IFC file was imported in CYPETHERM HE directly from Revit, all the thermal zones showed 0 m² and 75 m³. However, after exporting the IFC from the Open BIM Analytical Model and importing it into CYPETHERM HE, the thermal zone dimensions were not accurately transferred. The dining room thermal zone presented the biggest difference from Revit, as its volume was 180% bigger than in the BIM. The other zones, in turn, were about 6% to 12% bigger than in Revit.

Another problem found was the creation of two internal walls in the group of external walls that could not be deleted. In addition, the floor that was in contact with the ground was not transferred and, so, CYPETHERM HE considered this floor as a slab between floors, which could be a relevant issue when analysing the building energy demand. In contrast, all windows were properly imported, according to their dimensions, with only one window missing (see Figure 9).

Similarly to Open Studio, it is possible to edit the materials of the building envelope and to delete and add thermal zones in CYPETHERM HE. However, once a thermal zone or a building surface is added, it cannot be referred to as specific space and, so, it is pointless to make any change, in this aspect, in CYPETHERM HE. If there is necessity for modifying the model geometry or thermal zones, it can be done in the BIM authoring tool, or in another CYPE software called IFC Builder.

3.2.2. Warehouse Case in CYPETHERM HE

On the other hand, as happened at the moment of importing the warehouse IFC in Open BIM Analytical Model, not all the thermal zones were imported in CYPETHERM HE (see Figure 9). This might be a problem relating to extracting the data from the BIM authoring tool to the IFC file, once it worked properly with the gbXML format.

Although the construction of the model seemed accurate, the thermal zones were built with many errors. Firstly, some of the zones did not have all their enclosures or some surfaces presented gaps; furthermore, they did not have the correct dimensions, compared to the BIM. The thermal zone areas and volumes were bigger than in Revit; the same happened in the residential case. Moreover, as occurred in Design Builder, some surfaces of the thermal envelope were divided into many small parts, despite having the same constructive system.

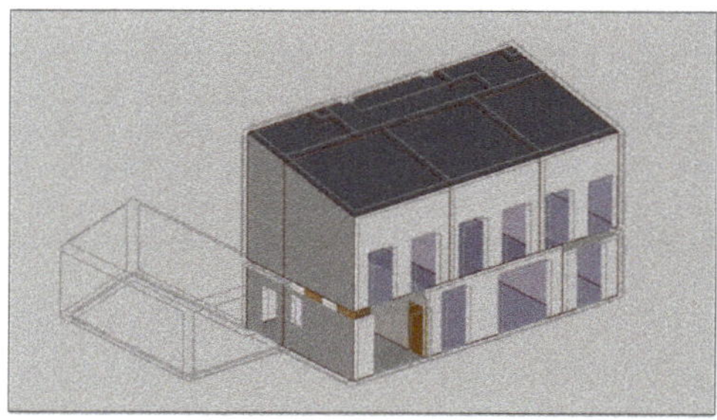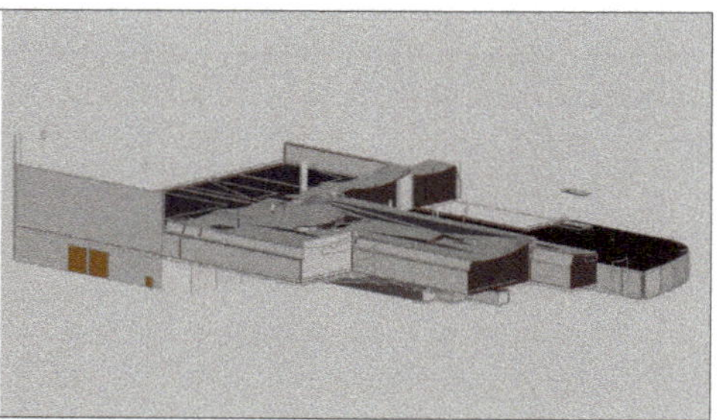

Figure 9. Residential and warehouse industry foundation classes (IFC) files imported in CYPETHERM HE.

Finally, as can be seen in Table 3, the problems detected in the IFC import in CYPETHERM HE increased with the size of the building and its complexity. In the residential case, CYPETHERM HE was able to verify each constructive system and detect if there were any missing elements. Nonetheless, the warehouse building had its surfaces divided into many parts; in some of which the construction systems were changed and, as consequence, it became difficult to measure and evaluate them. In addition, contrarily to Design Builder, in CYPETHERM HE, it is not possible to correct model issues. One way to solve these problems would be to import the IFC file into IFC Builder and to try to fix them there; however, this work would take a long time and might still not be totally free of error.

Table 3. Comparative table between IFC imported in CYPETHERM HE.

		CYPETHERM HE	
		IFC from Revit	
			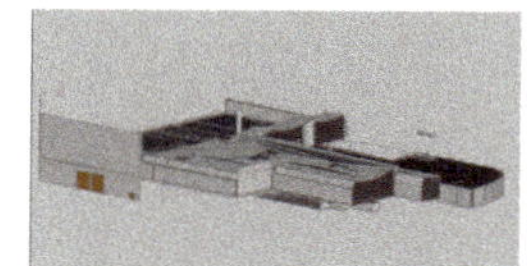
	The model	**Residential case**	**Warehouse case**
Dimensions	1. External dimensions.	Model between 0.70% to 7.50% smaller than in the BIM.	The dimensions were measured selecting the facades. Some of them were more than 20% bigger and others 10% smaller than in the BIM.
	1. Solution.	It is not possible to correct the external dimensions in CYPETHERM HE.	It is not possible to correct the external dimensions in CYPETHERM HE.
	2. Thermal zones area.	Areas between 6% to 12% bigger than in the BIM.	Areas between 1% to 5% bigger than in the BIM.
	2. Solution.	It is not possible to correct the thermal zone areas in CYPETHERM HE.	It is not possible to correct the thermal zone areas in CYPETHERM HE.
	3. Thermal zones volume.	Volumes between 12% to 134% bigger than in the BIM.	Volumes between 27% to 180% bigger than in the BIM.
	3. Solution.	It is not recommended to modify the thermal zone volumes in CYPETHERM HE.	It is not recommended to modify the thermal zone volumes in CYPETHERM HE.

Table 3. *Cont.*

		CYPETHERM HE	
Geometry	4. Zone recognition.	Correct. Total of 13 zones.	Incorrect. Total of 19 zones, not 21.
	4. Solution.	-	It is not possible to correct the zone recognition in CYPETHERM HE.
	5. Building surfaces.	Gaps in the surfaces of the thermal envelope. Some walls and the ground floor had the wrong recognition of their constructive system. One window was missing.	Missing walls in the corresponding thermal zone. Gaps in the surfaces of the thermal envelope. The roof was divided into many parts, although it had the same construction system. Some constructive systems did not have any building surface assigned.
	5. Solution.	It is not possible to correct the building geometry in CYPETHERM HE.	It is not possible to correct the building geometry in CYPETHERM HE.
Materials	6. Building materials.	The name of the constructive system was imported correctly, but none of the materials could be imported from Revit.	The name of the constructive system was imported correctly, but none of the materials could be imported from Revit.
	6. Solution.	Create the typologies in CYPETHERM HE.	Create the typologies in CYPETHERM HE.
	7. Material thickness.	None of the materials could be imported from Revit.	None of the materials could be imported from Revit.
	7. Solution.	Create the typologies in CYPETHERM HE.	Create the typologies in CYPETHERM HE.
	8. Material thermal properties.	None of the materials could be imported from Revit.	None of the materials could be imported from Revit.
	8. Solution.	Create the typologies in CYPETHERM HE.	Create the typologies in CYPETHERM HE.

3.3. Evaluation of the Processes

As this article aims to evaluate the interoperability of BIM to BEM, all the processes are summarized in Table 4, in order to determine which one is the most feasible to implement in a daily office routine, under the vision of an architect user of the softwares. The feasibility was analysed according to the time spent, the effort dedicated, and the difficulty presented in each process, rated on a scale from 1 to 5, where 1 represents the easiest process and 5 represents the most difficult. This analysis was made according to a generic model which had a complexity level similar to that of the warehouse case.

As can be seen from Table 4, exporting to the gbXML file was considered easier than the IFC format. At the moment of exporting the IFC from Revit, some specific parameters must be set correctly; otherwise, it cannot extract the data properly.

Table 4. Comparative table with assessment of the processes. Scale range: 1 to 5, where 1 means the most feasible.

Process	Export Revit gbXML	Export Revit IFC	Import Design Builder gbXML	Import Open Studio gbXML	CYPETHERM HE IFC
Export/Import the files	3	3	1	1	1
Verify the information transferred	-	-	4	4	3
Correct the dimension errors	-	-	3	5	5
Correct the geometry errors	-	-	3	5	5
Correct the material errors	-	-	3	3	2
Total	3	3	14	18	16

In turn, in the BEM softwares, importing the gbXML and IFC did not present any difficulty, as long as the interface at the time of import was correctly defined in the energy tool. The interoperability in Design Builder was considered the most feasible one and that in Open Studio was the least viable.

3.4. Simulation Results

After analyzing the import of the gbXML and IFC files of the residential building and the warehouse into BEM software, all the models were simulated with the aim of assessing whether the energy results would differ, with respect to the BEM software used. The simulations were executed monthly, from 1st January to 31st December, using the weather file of Madrid, Spain, under ideal loads.

However, among the three processes carried out, only the first and the third related to the residential case presented results. The gbXML file of the house imported in Open Studio showed some fatal errors in its geometry (e.g., missing walls and floors), which made its energy calculation impossible. Furthermore, the warehouse case was not simulated. Its gbXML and IFC files were imported in BEM tools without any problem; however, the model was not constructed properly. Missing elements and gaps in the thermal envelope, such as in the roof and external walls, were among the fatal errors encountered.

Therefore, only the results of the first and the third processes (see Figure 1) of the residential case were analysed:

1. gbXML file exported from Revit and imported in Design Builder.
2. IFC file exported from Revit and imported in CYPETHERM HE.

As this research intends to verify the interoperability between the data transfer from BIM to BEM, at the moment of comparing the results, two constructive elements were selected that can be inserted and modified in BIM and are essential to the study of energy modelling: the windows and the opaque envelope.

- Windows Heat Gain

As can be seen in Figure 10, there was a big difference between the model results of the residential case. First of all, the windows heat addition value of the first process was six times smaller than the third case, simulated in CYPETHERM HE. Once the huge discrepancy among these values was determined, we intended to discover where the problem originated from and, so, we collected the information about three parameters that could have influenced the window heat addition: Area of multiplied openings (m²), glass U-factor, and glass SHGC (Table 5).

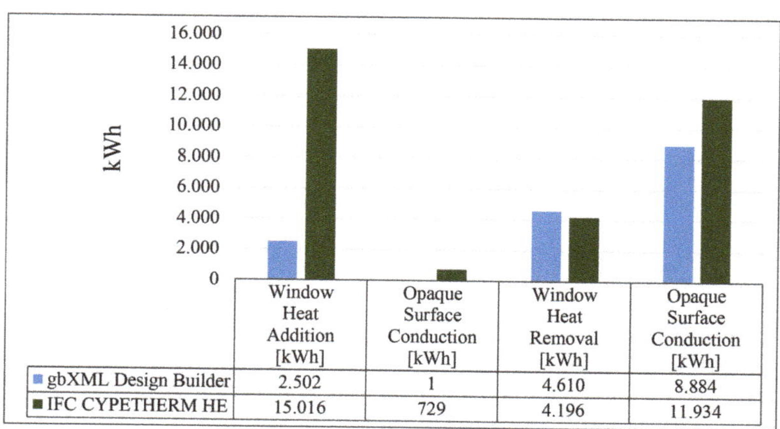

	Window Heat Addition [kWh]	Opaque Surface Conduction [kWh]	Window Heat Removal [kWh]	Opaque Surface Conduction [kWh]
gbXML Design Builder	2.502	1	4.610	8.884
IFC CYPETHERM HE	15.016	729	4.196	11.934

Figure 10. Heat gain components of the residential building simulations.

Table 5. Window properties.

	Windows Heat Addition	Area of Openings [m²]	Glass U-Factor [W/m²K]	Glass SHGC	Windows Frame [Total m²]
Revit	-	39.24	3.69	0.78	6.70
gbXML in Design Builder	2501.83	38.13	2.93	0.13	-
IFC in CYPETHERM HE	15,015.64	33.00	3.68	0.70	-

Therefore, we found that, in the gbXML file, the glass U-values and SHGC were not equal to those in Revit (see Table 5). For some reason, the U-value was 2925 W/m² K and the SHGC was 0.13, while, in Revit, they were 3.68 W/m² K and 0.78, respectively. This was a failure in the transfer of the data from BIM to the gbXML file, which explains the lowest value of window heat addition among the two cases.

On the other hand, the third process, which had the IFC file imported in the CYPETHERM HE software, presented the exact same glass U-value as in Revit. However, at the moment of analysing the IFC file in Notepad, we detected that the glass SHGC was not transferred to the IFC and, therefore, a SHGC default value equal to 0.70 was set automatically in CYPETHERM HE.

Regarding the area of openings, gbXML had a value closer to the BIM. In Revit, the window total was 39.24 m², while the gbXML and IFC files presented values 3% and 15% smaller than BIM, respectively.

Another interoperability problem was related to the window frames. In the BIM model, the openings had wooden frames, but their thermal properties were not defined. This information was not extracted to the gbXML and IFC files and, therefore, none of this opaque fraction of the opening was transferred to the BEM.

- Opaque Surface Conduction

As can be seen from Table 6, there were differences between the areas and U-values of the thermal envelope opaque elements (i.e., floors, walls, and roof).

Table 6. Opaque Surface Conduction Elements.

	Opaque Surface Conduction Addition [kWh]	Exterior Wall Area [m²]	Exterior Wall U-Factor [W/m²K]	Ground Floor Area [m²]	Ground Floor U-Factor [W/m²K]	Sloped Roof Area [m²]	Sloped Roof U-Factor [W/m²K]
Revit	-	309.00	0.33	113.00	3.03	98.00	0.41
gbXML in Design Builder	0.76	132.84	0.46	67.05	1.92	No information	0.44
IFC in CYPETHERM HE	728.50	243.05	0.37	No information	0.61	84.40	0.36

The gbXML presented the lowest value for the exterior walls: 132.84 m², 57% less than the area in the BIM (309.00 m²). In addition, the IFC also showed a small value, but only 21% less. Moreover, the U-factor in the first process was 39% bigger than in the BIM model baseline, which was inconsistent as, in Design Builder, a template was created with all Revit materials imported. Consequently, the material properties should have been the same.

In relation to the ground floor, clearly there were issues regarding the recognition of the imported data. First of all, in the simulation of the third case, there was no information about the area of this floor, and its U-value was much lower than in the BIM (by 80%). Moreover, the gbXML also showed differences with Revit, as its surface area was 40.67% smaller and its U-factor was 36.63% lower.

Finally, referring to the sloped roof, the U-factor values did not differ much from the baseline model. However, in the first simulation process, any information about the roof area was extracted and, in the IFC, there was a difference of −13.87%.

These differences between BIM and BEM must be taken into account, as they are relevant to the energy analysis. In addition, in a smaller scale building, it is easier to correct the differences in the thermal properties of the envelope materials. However, when it comes to a bigger building, the work required to correct them would take a long time and would be prone to human error, by missing some property that was transferred incorrectly. Therefore, it is crucial that these values are either the same as in the BIM authoring tool, or present very small differences that can be disregarded.

4. Conclusions

The lack of BIM–BEM interoperability is one of the existing gaps between digitalisation and the construction area. As BEMs are the basis of energy performance certificates (EPCs), they should present accurate results, in order to guarantee the confidence of investors in the energy efficiency sector.

It was observed that there exists a semi-automated workflow between BIM and BEM. First of all, at the moment of exporting the gbXML and IFC files, some parameters must be activated in Revit. Secondly, it is necessary to set the BEM tool appropriately for the model schema data import, with respect to the building typology, ideal loads, occupation schedule, and weather file.

However, their configuration was not sufficient to ensure adequate interoperability, as many of the problems encountered in the BEM softwares were consequences of errors from transfer of the BIM authoring tool data to gbXML and IFC files. Neither the energy models created with gbXML nor IFC can be trusted as a reference for the BIM, as they presented thermal properties and geometry values different from the baseline model, which led to distinctive and erroneous simulation results.

In addition, the BIM–BEM interoperability does not work for all types of buildings. It was shown that the bigger and more complex the building is (related to its shape and constructive systems), the less reliable the data transferred is and, so, the greater the problems in creating the model in the BEM software. These geometry issues resulted in fatal errors and made the energy simulations impossible.

In summary, BIM–BEM interoperability still has many shortcomings. We generated simulation results for the residential building, but they were not reliable. On the other hand, the warehouse could not even be simulated. For these reasons, it is possible to conclude that greater confidence can be had when recreating the model in a native BEM tool, in order to guarantee the better quality of results without geometric or construction component errors. Therefore, it is understood that the BIM to BEM interoperability is not ready to save time and costs in the design process, under the current state-of-the-art. A bigger effort should be made by standardization bodies, in order to increase the confidence of stakeholders.

Author Contributions: conceptualization, G.B.P., C.F.B., K.D.V.d.L. and A.S.-O.G.; methodology, G.B.P., C.F.B., K.D.V.d.L. and A.S.-O.G.; software, G.B.P. and K.D.V.d.L.; validation, G.B.P., C.F.B. and K.D.V.d.L.; investigation, G.B.P. and K.D.V.d.L.; resources, C.F.B.; writing—original draft preparation, G.B.P.; writing—review and editing, G.B.P., C.F.B., K.D.V.d.L. and A.S.-O.G.; supervision, C.F.B. and A.S.-O.G.; project administration, C.F.B.; funding acquisition, C.F.B. All authors have read and agreed to the published version of the manuscript.

Funding: This research was funded by the Government of Navarra (Spain) under the project "From BIM to BEM: B&B" (ref. 0011-1365-2020-000227).

Institutional Review Board Statement: Not applicable.

Informed Consent Statement: Not applicable.

Data Availability Statement: Not applicable.

Acknowledgments: We would like to thank Fundación Saltoki for making its warehouse BIM model available, which served as a case study for this research. We also thank CYPE Ingenieros for their support and collaboration during this study.

Conflicts of Interest: The authors declare no conflict of interest.

Abbreviations

The following abbreviations are used in this manuscript:

BIM	Building Information Modelling
BEM	Building Energy Model
EPC	Energy Performance Certificates
HTML	Hypertext Markup Language
XHTML	Extensible Hypertext Markup Language
bcXML	Building Construction Extensible Markup Language
IFCXML	Industry Foundation Classes Extensible Markup Language
IFC	Industry Foundation Classes
gbXML	Green Building Extensible Markup Language

References

1. Ruiz, G.R.; Bandera, C.F.; Temes, T.G.A.; Gutierrez, A.S.O. Genetic algorithm for building envelope calibration. *Appl. Energy* **2016**, *168*, 691–705. [CrossRef]
2. Hogeling, J.; Derjanecz, A. The 2nd recast of the Energy Performance of Buildings Directive (EPBD). *Eu Policy News Rehva J.* **2018**, *55*, 71–72.
3. Parliament, E. Directive 2018/844 of the European Parliament and of the Council of 30 May 2018 amending Directive 2010/31/EU on the energy performance of buildings and Directive 2012/27/EU on energy efficiency. *Off. J. Eur. Union* **2018**, *156*, 75–91.
4. Li, Y.; Kubicki, S.; Guerriero, A.; Rezgui, Y. Review of building energy performance certification schemes towards future improvement. *Renew. Sustain. Energy Rev.* **2019**, *113*, 109244. [CrossRef]
5. Asdrubali, F.; Baggio, P.; Prada, A.; Grazieschi, G.; Guattari, C. Dynamic life cycle assessment modelling of a NZEB building. *Energy* **2020**, *191*, 116489. [CrossRef]
6. González, V.G.; Ruiz, G.R.; Segarra, E.L.; Gordillo, G.C.; Bandera, C.F. Characterization of building foundation in building energy models. In Proceedings of the Building Simulation, Rome, Italy, 2–4 September 2019.
7. Elagiry, M.; Dugue, A.; Costa, A.; Decorme, R. Digitalization Tools for Energy-Efficient Renovations. *Proceedings* **2020**, *65*, 5009. [CrossRef]
8. Elagiry, M.; Marino, V.; Lasarte, N.; Elguezabal, P.; Messervey, T. BIM4Ren: Barriers to BIM implementation in renovation processes in the Italian market. *Buildings* **2019**, *9*, 200. [CrossRef]
9. Bergonzoni, G.; Marino, V.; Elagiry, M.; Costa, A. Assessing Residential Buildings Compliance with Sustainability Rating Systems through a BIM-Based Approach. *Proceedings* **2021**, *65*, 22. [CrossRef]
10. Ullah, K.; Lill, I.; Witt, E. An overview of BIM adoption in the construction industry: Benefits and barriers. In Proceedings of the 10th Nordic Conference on Construction Economics and Organization, Tallinn, Estonia, 7–8 May 2019; Emerald Publishing Limited: Bingley, UK, 2019.
11. Maskuriy, R.; Selamat, A.; Ali, K.N.; Maresova, P.; Krejcar, O. Industry 4.0 for the construction industry how ready is the industry? *Appl. Sci.* **2019**, *9*, 2819. [CrossRef]
12. de Lange, P.; Bähre, B.; Finetti-Imhof, C.; Klamma, R.; Koch, A.; Oppermann, L. Socio-technical Challenges in the Digital Gap between Building Information Modeling and Industry 4.0. In Proceedings of the 3rd International Workshop on Socio-Technical Perspective in IS Development (STPIS@ CAiSE), Essen, Germany, 13 June 2017, pp. 33–46.
13. González, V.G.; Colmenares, L.Á.; Fidalgo, J.F.L.; Ruiz, G.R.; Bandera, C.F. Uncertainy's Indices Assessment for Calibrated Energy Models. *Energies* **2019**, *12*, 2096. [CrossRef]
14. Fernández Bandera, C.; Ramos Ruiz, G. Towards a new generation of building envelope calibration. *Energies* **2017**, *10*, 2102. [CrossRef]
15. Dodier, R.H.; Henze, G.P. Statistical analysis of neural networks as applied to building energy prediction. *J. Sol. Energy Eng.* **2004**, *126*, 592–600. [CrossRef]
16. Reynders, G.; Diriken, J.; Saelens, D. Quality of grey-box models and identified parameters as function of the accuracy of input and observation signals. *Energy Build.* **2014**, *82*, 263–274. [CrossRef]
17. Ramos Ruiz, G.; Lucas Segarra, E.; Fernández Bandera, C. Model predictive control optimization via genetic algorithm using a detailed building energy model. *Energies* **2019**, *12*, 34. [CrossRef]
18. Gutiérrez González, V.; Ramos Ruiz, G.; Fernández Bandera, C. Empirical and Comparative Validation for a Building Energy Model Calibration Methodology. *Sensors* **2020**, *20*, 5003. [CrossRef]
19. Choi, J.; Shin, J.; Kim, M.; Kim, I. Development of openBIM-based energy analysis software to improve the interoperability of energy performance assessment. *Autom. Constr.* **2016**, *72*, 52–64. [CrossRef]
20. Abanda, F.; Byers, L. An investigation of the impact of building orientation on energy consumption in a domestic building using emerging BIM (Building Information Modelling). *Energy* **2016**, *97*, 517–527. [CrossRef]

21. Segarra, E.L.; Ruiz, G.R.; González, V.G.; Peppas, A.; Bandera, C.F. Impact Assessment for Building Energy Models Using Observed vs. Third-Party Weather Data Sets. *Sustainability* **2020**, *12*, 6788. [CrossRef]

22. Pezeshki, Z.; Soleimani, A.; Darabi, A. Application of BEM and using BIM database for BEM: A review. *J. Build. Eng.* **2019**, *23*, 1–17. [CrossRef]

23. Spiridigliozzi, G.; Pompei, L.; Cornaro, C.; De Santoli, L.; Bisegna, F. BIM-BEM support tools for early stages of zero-energy building design. *Iop Conf. Ser. Mater. Sci. Eng.* **2019**, *609*, 072075. [CrossRef]

24. Farzaneh, A.; Monfet, D.; Forgues, D. Review of using Building Information Modeling for building energy modeling during the design process. *J. Build. Eng.* **2019**, *23*, 127–135. [CrossRef]

25. Aranda, J.; Martin-Dorta, N.; Naya, F.; Conesa-Pastor, J.; Contero, M. Sustainability and Interoperability: An Economic Study on BIM Implementation by a Small Civil Engineering Firm. *Sustainability* **2020**, *12*. [CrossRef]

26. Pinheiro, S.; Wimmer, R.; O'onnell, J.; Muhic, S.; Bazjanac, V.; Maile, T.; Frisch, J.; van Treeck, C. MVD based information exchange between BIM and building energy performance simulation. *Autom. Constr.* **2018**, *90*, 91–103. [CrossRef]

27. Gerrish, T.; Ruikar, K.; Cook, M.; Johnson, M.; Phillip, M. Using BIM capabilities to improve existing building energy modelling practices. *Eng. Constr. Archit. Manag.* **2017**. [CrossRef]

28. Hijazi, M.; Kensek, K.; Konis, K. Bridging the gap: Supporting data transparency from BIM to BEM. *Archit. Res.* **2015**, *150*.

29. Sanhudo, L.; Ramos, N.M.; Martins, J.P.; Almeida, R.M.; Barreira, E.; Simões, M.L.; Cardoso, V. Building information modeling for energy retrofitting—A review. *Renew. Sustain. Energy Rev.* **2018**, *89*, 249–260. [CrossRef]

30. Kamel, E.; Memari, A.M. Review of BIM's application in energy simulation: Tools, issues, and solutions. *Autom. Constr.* **2019**, *97*, 164–180. [CrossRef]

31. Akbarieh, A. Systematic Investigation of Interoperability Issues between Building Information Modelling and Building Energy Modelling-BIM-Based BEM Information Exchange Issues. Master's Thesis, NTNU, Taipei, Taiwan, 2018.

32. Dimitriou, V.; Firth, S.K.; Hassan, T.M.; Fouchal, F. BIM enabled building energy modelling: Development and verification of a GBXML to IDF conversion method. In Proceedings of the 3rd IBPSA-England Conference BSO, Great North Museum, Newcastle, 12–14 September 2016; p. 1126.

33. Wetter, M.; van Treeck, C.; Helsen, L.; Maccarini, A.; Saelens, D.; Robinson, D.; Schweiger, G. IBPSA Project 1: BIM/GIS and Modelica framework for building and community energy system design and operation–ongoing developments, lessons learned and challenges. In Proceedings of the IOP Conference Series: Earth and Environmental Science, Graz, Austria, 11–14 September 2019; IOP Publishing: Bristol, UK, 2019; Volume 323, p. 012114.

34. Elagiry, M.; Charbel, N.; Bourreau, P.; Di Angelis, E.; Costa, A. IFC to Building Energy Performance Simulation: A Systematic Review of the Main Adopted Tools and Approaches. In Proceedings of the BauSIM 2020-8th Conference of IBPSA Germany and AustriaAt, Graz University of Technology, Graz, Austria, 23–25 September 2020.

35. Gumbarević, S.; Burcar Dunović, I.; Milovanović, B.; Gaši, M. Method for Building Information Modeling Supported Project Control of Nearly Zero-Energy Building Delivery. *Energies* **2020**, *13*, 5519. [CrossRef]

36. Gao, H.; Koch, C.; Wu, Y. Building information modelling based building energy modelling: A review. *Appl. Energy* **2019**, *238*, 320–343. [CrossRef]

37. Osello, A.; Cangialosi, G.; Dalmasso, D.; Di Paolo, A.; Turco, M.L.; Piumatti, P.; Vozzola, M. Architecture data and energy efficiency simulations: BIM and interoperability standards. In Proceedings of the Building Simulation, Sydney, Australia, 14–16 November 2011.

38. Crawley, D.B.; Lawrie, L.K.; Winkelmann, F.C.; Buhl, W.F.; Huang, Y.J.; Pedersen, C.O.; Strand, R.K.; Liesen, R.J.; Fisher, D.E.; Witte, M.J.; et al. EnergyPlus: Creating a new-generation building energy simulation program. *Energy Build.* **2001**, *33*, 319–331. [CrossRef]

39. Zhang, L. Simulation analysis of built environment based on design builder software. *Appl. Mech. Mater. Trans. Tech. Publ.* **2014**, *580*, 3134–3137. [CrossRef]

40. Guglielmetti, R.; Macumber, D.; Long, N. *OpenStudio: An Open Source Integrated Analysis Platform*; Technical Report; National Renewable Energy Lab. (NREL): Golden, CO, USA, 2011.

41. Ingenieros, C. Software Para Ingenieria, Arquitectura y Construccion. Available online: http://www.cype.es (accessed on 31 January 2021).

42. van Treeck, C.; Wimmer, R.; Maile, T. BIM for energy analysis. In *Building Information Modeling*; Springer: Berlin/Heidelberg, Germany, 2018; pp. 337–347.

43. Alsharif, R. A Review on the Challenges of BIM-Based BEM Automated Application in AEC Industry. 2019. Available online: https://www.researchgate.net/publication/337888446_A_review_on_the_challenges_of_BIM-based_BEM_automated_application_in_AEC_industry (accessed on 12 February 2021).

44. Moretti, N.; Xie, X.; Merino, J.; Brazauskas, J.; Parlikad, A.K. An openBIM Approach to IoT Integration with Incomplete As-Built Data. *Appl. Sci.* **2020**, *10*, 8287. [CrossRef]

applied sciences

Article

Accurate Prediction of Hourly Energy Consumption in a Residential Building Based on the Occupancy Rate Using Machine Learning Approaches

Le Hoai My Truong [1], Ka Ho Karl Chow [1], Rungsimun Luevisadpaibul [1], Gokul Sidarth Thirunavukkarasu [1,*], Mehdi Seyedmahmoudian [1,*], Ben Horan [2], Saad Mekhilef [1,3] and Alex Stojcevski [1]

[1] School of Software and Electrical Engineering, Swinburne University of Technology, Hawthorn, VIC 3122, Australia; 102083456@student.swin.edu.au (L.H.M.T.); 101930616@student.swin.edu.au (K.H.K.C.); 101584262@student.swin.edu.au (R.L.); saad@um.edu.my (S.M.); astojcevski@swin.edu.au (A.S.)

[2] School of Engineering, Deakin University, Waurn Ponds, VIC 3216, Australia; ben.horan@deakin.edu.au

[3] Department of Electrical of Engineering, University of Malaya, Kuala Lumpur 50603, Malaysia

* Correspondence: gthirunavukkarasu@swin.edu.au (G.S.T.); mseyedmahmoudian@swin.edu.au (M.S.)

Citation: Truong, L.H.M.; Chow, K.H.K.; Luevisadpaibul, R.; Thirunavukkarasu, G.S.; Seyedmahmoudian, M.; Horan, B.; Mekhilef, S.; Stojcevski, A. Accurate Prediction of Hourly Energy Consumption in a Residential Building Based on the Occupancy Rate Using Machine Learning Approaches. *Appl. Sci.* **2021**, *11*, 2229. https://doi.org/10.3390/app11052229

Academic Editor: Chiara Bedon

Received: 2 February 2021
Accepted: 25 February 2021
Published: 3 March 2021

Publisher's Note: MDPI stays neutral with regard to jurisdictional claims in published maps and institutional affiliations.

Abstract: In this paper, a novel deep neural network-based energy prediction algorithm for accurately forecasting the day-ahead hourly energy consumption profile of a residential building considering occupancy rate is proposed. Accurate estimation of residential load profiles helps energy providers and utility companies develop an optimal generation schedule to address the demand. Initially, a comprehensive multi-criteria analysis of different machine learning approaches used in energy consumption predictions was carried out. Later, a predictive micro-grid model was formulated to synthetically generate the stochastic load profiles considering occupancy rate as the critical input. Finally, the synthetically generated data were used to train the proposed eight-layer deep neural network-based model and evaluated using root mean square error and coefficient of determination as metrics. Observations from the results indicated that the proposed energy prediction algorithm yielded a coefficient of determination of 97.5% and a significantly low root mean square error of 111 Watts, thereby outperforming the other baseline approaches, such as extreme gradient boost, multiple linear regression, and simple/shallow artificial neural network.

Keywords: deep learning; energy management systems; load forecasting; machine learning and microgrids

1. Introduction

The Australian energy market has been operating on a centralised generation model with state-owned power plants situated closest to fossil fuel resources such as coal, hydro, wind, and natural gas for many years. The centralised electricity generation model demonstrates several drawbacks to the environment and end-users due to the reduced efficiency caused by large transmission losses. The electricity price keeps going up due to the increased investment in distribution infrastructure required to connect households and businesses to a stabilised power supply [1]. Therefore, the government has spent the last couple of decades shifting its electricity generation model to a more decentralised approach by incorporating more renewable resources. Despite the tremendous amount of investment, there are still many transitional challenges to be solved, both in terms of political and technological readiness. It is predicted that the role of grid-supplied power will be inverted from being the primary source of energy to a backup source, having distributed renewable generation as the primary source [1]. This paper investigates a similar distributed generation model commonly known as microgrids and addresses the systems' technical barrier of accurate forecasting of load demand, aiming to enable broader adoption of the decentralised generation model.

121

The microgrid consists of a control unit that uses a robust energy management system equipped with advanced load demand/generation capacity forecasting algorithms that aids in achieving reliable power flow control, load sharing, grid protection, stability, and smooth operation [2]. If successfully implemented, micro-grid systems promise to offer distribution system congestion relief, the postponement of new generation or delivery capacity, response to load changes, and local voltage support [2]. Accurate forecasting of load demand is crucial to a microgrid's energy management system because it ensures energy savings and improves the operational or sizing efficiency of its supply and storage. However, load forecasting can be a tricky task since the energy performance of a building is influenced by various factors, such as occupancy rate, residence behaviour, household income, number and type of appliances, and weather conditions.

This study aims to develop a novel deep neural network-based energy prediction algorithm to synthesise hourly load profiles based on the occupancy rate. To achieve this, a comprehensive multi-criteria analysis (MCA) was carried out on the existing literature initially, and a novel synthetic load profile generator was formulated. The MCA aims to identify the most suitable prediction algorithm by evaluating the accuracy, usability, adaptability, computation time, randomisation, and implementation feasibility. The synthetic load profile generator developed in this study is later used to generate the stochastic load profiles. A state-of-the-art predictive microgrid model was created to generate the load profile accurately using average load consumption data, occupancy rate, and seasonality as the input. The microgrid's mathematical model consists of realistic appliances with predefined constraints that help synthetically generate the accurate load profile. The proposed state-of-art eight-layered deep neural network model accurately predicts the hourly energy consumption patterns, considering the occupancy rate and seasonality as the critical inputs. Baseline models such as extreme gradient boost, multiple linear regression, and simple/shallow artificial neural network are used to verify the proposed prediction algorithm's performance. Root mean square error (RMSE) and coefficient of determination (R^2) values are the metrics that the models are compared on the results indicated that the proposed deep neural network model outperformed the other models.

The manuscript is structured as follows: A detailed background study on micro-grids and their energy management systems (EMS), followed by the foundation of different machine learning approaches used in load profile forecasting, is highlighted in Section 2. In Section 3, the study's methodology is deliberately discussed by highlighting the process of MCA table construction, the synthetic load profiles' generator model, and the assumptions considered. A detailed overview of the proposed model and the baseline approaches considered for evaluation is also discussed. In Section 4, the results and discussions are highlighted. Finally, the concluding remarks and the significance of the paper are highlighted in Section 5.

2. Background

2.1. Microgrid and Energy Management System

The decentralised electricity generation model has many advantages over the conventional centralised systems that consist of remote generation units. In a centralised generation model, power flows in one direction, from a small number of large generators to many consumers over a long distance. Therefore, the centralised model requires large power plants to meet the demand and significant transmission lines to connect households and businesses with their power source, resulting in colossal air pollutant emissions, wastage of generation, and land use. The fact that it requires a high integration level also means that its system is extremely vulnerable to disturbances in the supply chain. Therefore, its attractiveness is reducing, and the penetration of small-scale decentralised systems or microgrids is emerging and increasingly invested.

The microgrid is essentially a local energy grid with control capability and autonomous operation, which can be disconnected from the primary grid when required [3]. In opposition to centralised systems, power in a microgrid flows in both directions. As it is built locally using renewable sources, it is more efficient, more reliable, lower cost, and cleaner than the centralised model. The three main types of a microgrid are remote, grid-connected, and networked microgrids. In this study, a grid-connected microgrid focuses on the most suitable system for commercial and residential buildings. A grid-connected microgrid can operate in either grid-connected mode or stand-alone mode based on the requirement. In grid-connected mode, the microgrid imports or exports the power from and to the grid depending on generation and load conditions and market policies based on contractual obligations. On the other hand, in islanding mode, the microgrid disconnects from the utility when an abnormal condition occurs in the grid, and the microgrid has to satisfy the load with the required level of power quality by utilising the local storage and renewable resources [4]. Therefore, the energy management strategies used are different, and a coordinated control approach for microgrid energy management is required to minimise the errors between the forecasting and real-time data in the schedule, and dispatch layers of the system [5].

The setup of monitoring and optimising energy consumption to regulate the energy flow is commonly known as an energy management system (EMS). A microgrid usually requires an energy management system to assign active and reactive power references, ensuring cooperation between the controllable units to achieve stable and economic operation [5]. The latest research shows that the objective of energy management of a microgrid is to minimise the microgrid's operating costs such as fuel costs, operating maintenance costs, and purchase cost of electricity from the conventional power grids [5]. One of the key features that allow a microgrid energy management system to achieve its objective is to have a robust and accurate forecasting technique of load demand/generation profile capability. In a residential setting, energy management works on optimising energy consumption, equipment efficiency, detecting faults in a deteriorating system, implementing ways to reduce energy wastage, and recovering energy wastage for other purposes. The EMS inferences provide data visualisation to household owners to better understand their usage patterns and recommendations on smart energy usage to drive better behaviour in their daily use. As an effect, maintenance cost is reduced because equipment usage is optimised [6]. Not only capturing and presenting historical data, but an EMS can also help forecast a household or a building's energy consumption by using an intelligent machine learning algorithm. It also allows the developers to accurately identify the sizing of power resources required to meet a built or a new unit's demand. However, one of the main drawbacks of using high-end machine learning algorithms is the lack of valid data set for effective training. That is why our study focuses not only on developing an optimised machine learning algorithm but also on establishing a novel synthetic load generator to evaluate our proposed model.

For the synthetic load generator to be built, an investigation into standard controllable loads, uncontrollable loads, and critical loads with their usage pattern in a typical household was carried out. Any device or appliance's power consumption can be controlled by adjusting their duration on and off time [7]. Controllable loads can be differentiated from uncontrollable loads based on their ability to be turned off without sacrificing the user's comfort [8]. For example, air conditioners and refrigerators can be turned off for a certain period to save power without significantly affecting household comfort. However, microwave ovens or toasters, examples of uncontrollable load, should not be turned off because this directly impacts the occupant's comfort. The critical loads are loads that will result in a significant loss or damage when power off [9]. Examples of essential loads are a smoke alarm, a water pump, and critical load panels in an energy storage system [10]. Following these definitions, in a distributed power system, popular controllable loads, such as refrigerators, HVAC systems, entertainment devices, and fans are considered in the mathematical model formulated in this study. The most commonly used loads in residential settings are supposed to generate an extensive data set of load profiles with artificially induced non-linearity. A list of loads being considered is explained in the methodology section below.

Another fact about microgrids is that majority of them are built-in integration with renewable energy sources such as wind turbines, solar photovoltaic (PV), and fuel cells due to the global transition to green technologies. In addition to the eco-friendly advantages of such systems, the renewable resources add more non-linearity into the system due to their stochastic nature, resulting in an advanced/robust energy management algorithm. In 2009, for the EU, nearly 55% of the new installed capacity based on renewable sources corresponded to the wind and PV intermittent generation (39% and 16%, respectively) [11]. According to the Australian Department of Industry, Science, Energy, and Resources, in 2019, 21% of Australia's total electricity generation was based on the renewable energy resources, which consisted of 7% of wind, 7% of solar, and 5% of hydro, making the share of renewables the highest recorded since the 1970s [12]. On this note, the study of load forecasting methodology is critical because it will enable an accurate sizing/scheduling of renewable sources for a microgrid system, preventing resource wastage or shortage and many other system failures.

2.2. Machine Learning

The complexity of residential load forecasting lies in the significant volatility and stochastic nature of the load profiles. Many researchers worldwide are working towards addressing this complexity by developing an accurate forecasting technique that addresses this uncertainty. The statistical learning approaches are based on the predefined relationship between variables and require a smaller data set but whereas the more accurate and advanced machine learning algorithms require a big data set. In recent years, the rise of big data with machine learning makes it a potential solution to address load forecasting in a residential energy management system. Traditional methods tend to avoid such uncertainty by load aggregation to offset uncertainties, customer classification to cluster uncertainties, and spectral analysis to filter out uncertainties [13]. Therefore, many studies are carried out to evolve the current machine learning techniques to learn the uncertainty at the building level directly due to the many influencing factors.

According to Lars Hulstaert, a data scientist at Johnson and Johnson, most machine learning systems require the ability to explain to stakeholders why specific predictions are made [14]. The accuracy and interpretability trade-off is typically considered when choosing a suitable machine learning model. Generally, there are two types of machine learning models, namely, black-box and white-box. Black-box models such as neural networks and gradient boosting models yield highly accurate predictions. However, their computational operation is difficult to understand. On the other hand, white-box models such as linear regression and decision trees, despite being much easier to interpret, produce less predictive capacity. In this research, an initial comprehensive multi-criteria analysis of the most common machine learning techniques in models such as linear regression, gradient boosting, decision tree, and neural network was performed to determine the best potential method optimised for load forecasting application.

Machine learning approaches are generally used to address supervised and unsupervised learning problems. Since the proposed methodology aims to predict energy consumption, supervised learning is a more suitable option as its primary function is to model the value of the target variable based on the predictor variables. Machine learning and artificial intelligence techniques are used in a wide variety of applications, such as load forecasting [15], determining product quality [16], and fault quality [17]. Linear regression, decision trees, and neural networks are all examples of supervised learning. Despite the similarities, their computation principles are different. Regression analysis is a methodology that allows finding a functional relationship among dependent variables and independent variables [18]. For complex systems, such as the energy consumption in buildings, the regression analysis is considered as an iterative process, in which the outputs are used to diagnose, validate, criticise, and possibly modify the inputs [18]. In the decision tree approaches, an empirical tree represents a segmentation of the created data by applying a series of simple rules. Through the repetitive process of splitting, predictions are made, and the logic is usually comprehensible [19].

On the contrary, the neural network is a class of algorithms loosely modelled on connections between neurons in the human biological brain, which is designed to imitate the natural nervous system information process and decision making [20]. There are many choices of neural network optimising architecture that significantly influences the performance of the model. This study proposes a novel deep neural network model by optimising the hyper-parameters to enhance neural networks' performances. A comprehensive literature review in the form of multi-criteria analysis (MCA) was carried out. The next section of the paper will critically highlight how the MCA analysis is performed and evaluated.

3. Multi-Criteria Analysis (MCA)

3.1. MCA Development

A MCA to choose the most suitable machine learning technique for estimating energy consumption in a residential building is an evaluation process that considers different measurable criteria to rank, compare, and select the best performing models considered in the literature. A list of benchmarks is identified to evaluate the identified techniques' performance and measured either qualitatively or quantitatively. The MCA was established by following the procedure shown in Figure 1, the set of chosen criteria are listed in Table 1.

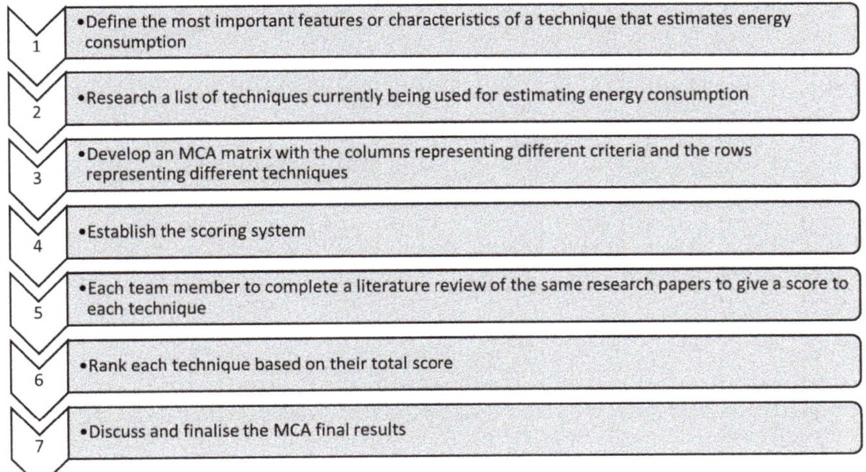

Figure 1. Procedures for establishing multi-criteria analysis (MCA).

Table 1. The set of chosen criteria for the MCA.

Implementation Feasibility	The level of ease to implement the technique in the restricted amount of time and resources.
Usability	The capacity to provide a condition for its users to perform the tasks safely, effectively, efficiently and satisfactorily.
Computational time	The amount of time it takes for the technique to converge to an outcome.
Accuracy	The size of the dispute between the technique's outcome and the real statistic.
Randomisation	The ability of the technique to draw a pattern from a random dataset.
Adaptability	The same technique can be combined with other optimisation technique or can be used in a different environment.

There are six main analysis criteria considered in this study, each with different weighting depending on their relative importance to the study's objectives. In this study, accuracy is assumed to have the highest weightage because it aims to identify the most accurate predictive method for estimating residential buildings' energy consumption. A rating from 1 to 10 is applied to each criterion, with a higher value representing a favourable outcome. Each technique will be ranked according to its MCA score, where the higher the score, the more suitable the approach fits for purpose. Furthermore, the MCA was done with three separate sets of scoring systems for different perspectives of business managers, electrical engineers, and data scientists to ensure the final result is not biased to one specific area. The three scoring systems used the same papers from the literature review, with the final scores being the average of the individual scores.

3.2. MCA Results Evaluation

The MCA matrices and final scoring table obtained from the study are indicated in Figure 2 and Table 2. The MCA consisted of eight different approaches used in load forecasting from the existing literature that were critically analysed. By considering different perspectives and analysing different performance criteria, the framework yielded an accurate shortlist of the most effective ML techniques in estimating energy consumption in a residential building.

Criteria Description	Implementation Feasibility	Usability	Computation time	Accuracy	Randomisation	Adaptability	Total score
	The level of ease to implement the technique in the restricted amount of time and resources.	The capacity to provide a condition for its users to perform the tasks safely, effectively, efficiently and satisfactorily.	The amount of time it takes for the technique to converge to an outcome.	The size of dispute between the technique's outcome and the real statistic.	The ability of the technique to draw pattern from a random dataset/ The randomness of data source.	The same technique can be combined with other optimisation technique or can be used in different environment.	
Weighting	4	4	4	5	3	3	
Business Manager							
ANFIS (Adaptive Neural Fuzzy Inference System)	7	8	9	9	9	9	195
ANN (Artificial Neural Network)	8	8	8	9	9	8	192
MLR (Multiple Linear Regression)	9	9	8	6	5	8	173
XGBoost (eXtreme Gradient Boosting)	9	9	9	8	7	7	190
WNN (Wavelet Neural Network)	9	9	8	9	8	9	200
SVM (Support Vector Machine)	8	8	6	8	8	8	176
ARIMA (Auto Regressive Integrated Moving Average)	8	8	9	7	5	8	174
Gaussian Process Regression	8	8	5	6	9	5	156
Electrical Engineer							
ANFIS (Adaptive Neural Fuzzy Inference System)	6	8	9	9	9	6	182
ANN (Artificial Neural Network)	5	8	7	9	9	9	179
MLR (Multiple Linear Regression)	10	9	9	6	5	9	184
XGBoost (eXtreme Gradient Boosting)	10	9	9	8	4	8	188
WNN (Wavelet Neural Network)	3	8	4	9	8	8	153
SVM (Support Vector Machine)	7	5	5	8	8	9	159
ARIMA (Auto Regressive Integrated Moving Average)	8	7	8	3	2	2	119
Gaussian Process Regression	7	4	6	7	9	4	142
Data Scientist							
ANFIS (Adaptive Neural Fuzzy Inference System)	9	9	8	7	7	8	184
ANN (Artificial Neural Network)	9	9	8	8	9	8	195
MLR (Multiple Linear Regression)	8	6	9	4	6	8	154
XGBoost (eXtreme Gradient Boosting)	9	7	8	7	7	8	176
WNN (Wavelet Neural Network)	7	9	8	9	9	7	189
SVM (Support Vector Machine)	8	7	8	7	8	7	172
ARIMA (Auto Regressive Integrated Moving Average)	7	6	7	4	4	5	127
Gaussian Process Regression	6	6	5	5	9	5	135

Figure 2. MCA matrices and scoring table.

Table 2. MCA total scoring table.

Techniques	Total Score
ANFIS (Adaptive Neural Fuzzy Inference System)	187
ANN (Artificial Neural Network)	189
MLR (Multiple Linear Regression)	170
XGBoost (eXtreme Gradient Boosting)	185
WNN (Wavelet Neural Network)	181
SVM (Support Vector Machine)	169
ARIMA (Auto Regressive Integrated Moving Average)	140
GPR (Gaussian Process Regression)	144

As the results, it was demonstrated that ML techniques based on neural networks such as ANN (MCA score = 189), ANFIS (MCA score = 187), and WNN (MCA score = 181) exhibited better performance in estimating energy consumption than those based on decision trees such as XGBoost (MCA score = 185) and regression analysis such as Gaussian process regression (MCA score = 144) and ARIMA (MCA score = 140). ANN produced the most accurate predicting results as it is well-known for its ability to handle noise and perform non-linear analysis of data-set like the investigated load profiles [21]. Furthermore, ANN also tends to ignore excess input data that are of minimal significance and concentrate on the more critical input [21]. On the other hand, despite performing better than MLR, XGBoost or the decision tree method generally does not outperform neural networks for non-linear data and is susceptible to noisy data [19]. The technique is more suitable for predicting categorical outcomes, and unless visible trends and sequential patterns are general, decision trees are less appropriate for application to time series data [19]. Regarding the MLR, despite being the most comfortable and most intuitive approach of prediction, it is the least appropriate for predicting energy usage due to its weakness in working with data with no apparent pattern.

Additionally, more criteria can be included for analysis to cover all the aspects of each ML technique. As a next step of the MCA, we model the performance of ANN, XGBoost, and MLR to validate the literature review on the performance of neural networks, decision tree, and regression analysis using python. The shallow or a simple ANN model is also used as a benchmark for testing the proposed prediction model's performance. Furthermore, a new hyperparameter-tuned deep neural network model will be developed and evaluated based on the prediction of energy consumption load profiles.

4. Methodology

The methodology adopted to rationalise the approach used in the study to maintain focus on critical research aspects is clearly illustrated in Figure 3.

The detailed procedures and assumptions of the load profiles synthesis and machine learning modelling are critically explained in the subsections below.

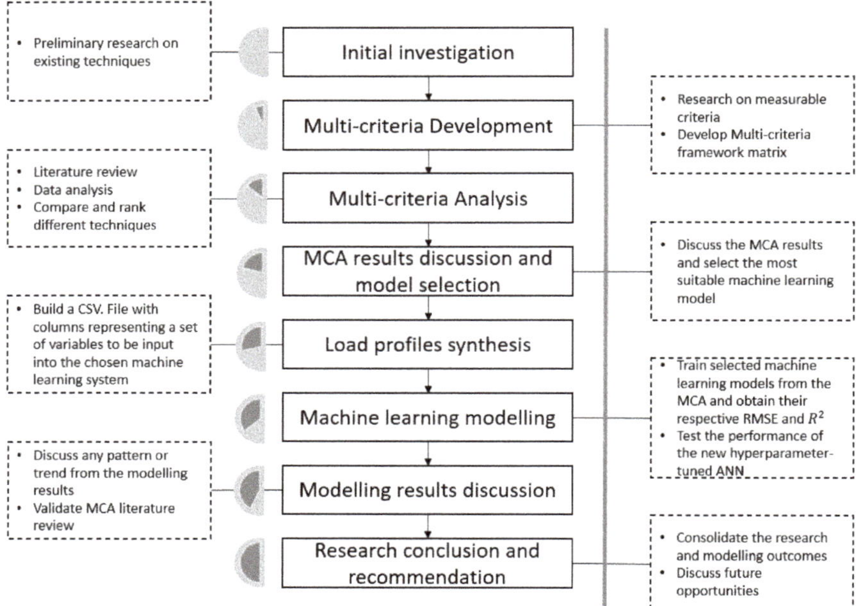

Figure 3. Research methodology.

4.1. Synthetic Load Profile Generation

In order to obtain a wide variety of load profiles based on the occupancy rate and seasonality, we developed a synthetic load profile generator that considers basic mathematical models of individual appliances to generate a set of load profiles programmatically. Initially, the model includes a load profile of different households with varying factors, such as the type of residence and number of people, as shown in Tables 3–8. Figure 4 provided a visual representation of the individual load profiles considered in the study. The synthetically generated household load profile models consist of a randomised algorithm that creates different appliance loads and the usage time. The load profiles are constrained to schedule the total amount of power consumption in kWh randomly and the electricity cost per day of the load profile to closely resemble real-world usage. The synthetic load profile generator is later used to generate a data sample of 100,000 data points (load profiles) for each type of occupancy rate, and these data are used by the proposed forecasting model to train and make predictions based on the limited number of inputs. The mathematical models are provided with a pre-defined set of constraints to generate the different non-linear load profiles replicating the real-time load profiles. Inferences from the power consumption studies on households are used as the basic input for the models to systematically generate numerous number of random samples of load profiles. These generated load profiles are then used to train a novel prediction algorithm to forecast the load profile at higher accuracy. The following factors are taken into consideration:

- The hourly electricity usage data used in the study were obtained from residential buildings with a different number of occupants in Victoria, Australia.
- Electricity bills and statistics provided by energy providers and distributors were used to identify the average daily usage, which was then considered as a reference point to fine-tune the load profiles to represent the daily usage in Victoria, Australia in 2020, and it was programmatically generated using the synthetic load profile generator.

- The power ratings for all of the appliances present in the load profile table were taken from the existing appliances and from the Daftlogic website, which provides the typical power consumption list of households [22].

Table 3. Initial load profile 1 created for the project (summer).

Type of Household	No. of Occupants	Load	Wattage (W)	Amount of Time per Day (h)	Prefer Usage Time	Energy Consumption (Wh)	Cost per Day (AUD)
Household 1: University student	1	Fridge	80	24	00:00–23:59	1920	
		Washing machine	800	0.1	10:00–11:00	80	
		Fan	200	10	13:00–23:00	2000	
		Microwave	800	0.1	12:00–13:00	80	
		Heater	2000	0	-	-	
		Chargeable devices	150	8	22:00–6:00	1200	
		Rice cooker	830	0.5	11:00–12:00	415	
		Toaster	850	0.05	7:00–8:00	42.5	
		TV	150	4	18:00–22:00	600	
		Gaming console	150	4	18:00–22:00	600	
		Other chargeable devices	200	6	10:00–16:00	1200	
	Total			Average S = 7.5, W = 10.4 kWh		8137.5	3.2924425

Table 4. Initial load profile 2 created for the project (summer).

Type of Household	No. of Occupants	Load	Wattage (W)	Amount of Time per Day (h)	Prefer Usage Time	Energy Consumption (Wh)	Cost per Day (AUD)
Household 2: 20 s couple	2	Fridge	120	24	00:00–23:59	2880	
		Washing machine	800	0.3	13:00–14:00	240	
		Fan	200	10	17:00–3:00	2000	
		Microwave	1000	0.2	12:00–13:00	100	
		Heater	2000	0	-	-	
		Chargeable devices	400	8	22:00–6:00	3200	
		Blender	500	0.1	9:00–10:00	50	
		Toaster	850	0.1	6:00–7:00	85	
		Iron	1200	0.1	19:00–20:00	120	
		Vacuum cleaner	1000	0.2	16:00–17:00	200	
		Gaming console	300	2	19:00–21:00	600	
		Coffee machine	1000	0.1	6:00–7:00	100	
	Total			Average S = 11.5, W = 14.6 kWh		9675	3.669745

Table 5. Initial load profile 3 created for the project (summer).

Type of Household	No. of Occupants	Load	Wattage (W)	Amount of Time per Day (h)	Prefer Usage Time	Energy Consumption (Wh)	Cost per Day (AUD)
Household 3:	3	Fridge	120	24	00:00–23:59	2880	
Young family		Washing machine	800	0.7	9:00–10:00	560	
		Fan	200	12	12:00–0:00	2400	
		Microwave	1000	0.5	11:00–12:00	500	
		Heater	2000	0	-	-	
		Chargeable devices	300	12	16:00–4:00	3600	
		TV	150	6	13:00–19:00	900	
		Rice cooker	850	1	11:00–12:00	850	
		Toaster	850	0.3	4:00–5:00	225	
		Iron	1200	0.5	20:00–21:00	600	
		Vacuum cleaner	1000	0.2	14:00–15:00	200	
		Dish washer	1500	1	18:00–19:00	1500	
		Gaming console	300	2	17:00–19:00	600	
		Coffee machine	1000	0.3	4:00–5:00	300	
		Total		Average S = 13.1, W = 17.6 kWh		15,345	5.061163

Table 6. Initial load profile 4 created for the project (summer).

Type of Household	No. of Occupants	Load	Wattage (W)	Amount of Time per Day (h)	Prefer Usage Time	Energy Consumption (Wh)	Cost per Day (AUD)
Household 4:	2	Fridge	120	24	00:00–23:59	2880	
Middle age couple		Washing machine	800	0.3	19:00–20:00	240	
		Fan	200	4	20:00–0:00	800	
		Microwave	1000	0.05	17:00–18:00	50	
		Heater	2000	0	-	-	
		Chargeable devices	200	12	18:00–6:00	2400	
		TV	150	3	20:00–23:00	450	
		Rice cooker	830	0.2	17:00–18:00	166	
		Toaster	850	0.1	5:00–6:00	85	
		Iron	1200	0.2	19:00–20:00	240	
		Vacuum cleaner	1000	0.1	16:00–17:00	100	
		Gaming console	150	1	20:00–21:00	150	
		Coffee machine	1000	0.2	5:00–6:00	200	
		Total		Average S = 11.5, W = 14.6 kWh		10,461	3.8626294

Table 7. Initial load profile 5 created for the project (summer).

Type of Household	No. of Occupants	Load	Wattage (W)	Amount of Time per Day (h)	Prefer Usage Time	Energy Con-sumption (Wh)	Cost per Day (AUD)
Household 5:	4	Fridge	150	24	00:00–23:59	3600	
Family		Washing machine	800	0.7	9:00–10:00	560	
		Fan	200	10	8:00–18:00	2000	
		Microwave	1000	0.5	10:00–11:00	500	
		Heater	2000	0	-	-	
		Chargeable devices	300	12	20:00–8:00	3600	
		TV	150	3	19:00–22:00	450	
		Toaster	850	0.3	6:00–7:00	255	
		Iron	1200	0.5	6:00–7:00	600	
		Vacuum cleaner	1000	0.2	11:00–12:00	200	
		Dish washer	1500	1	18:00–19:00	1500	
		Air fryer	1000	0.2	11:00–12:00	200	
		Gaming console	150	2	20:00–22:00	300	
		Coffee machine	1000	0.3	6:00–7:00	300	
		Total		Average S = 14.5, W = 18.9 kWh		15,215	5.029261

Table 8. Initial load profile 6 created for the project (summer).

Type of Household	No. of Occupants	Load	Wattage (W)	Amount of Time per Day (h)	Prefer Usage Time	Energy Con-sumption (Wh)	Cost per Day (AUD)
Household 6:	5	Fridge	180	24	00:00–23:59	4320	
Big family		Washing machine	800	0.7	12:00–13:00	560	
		Fan	200	12	12:00–24:00	2400	
		Microwave	1000	0.5	11:00–12:00	500	
		Heater	2000	0	-	-	
		Chargeable devices	300	12	20:00–8:00	3600	
		TV	150	6	16:00–22:00	900	
		Rice cooker	850	1	11:00–12:00	850	
		Toaster	850	0.3	6:00–7:00	255	
		Iron	1200	0.5	19:00–20:00	600	
		Vacuum cleaner	1000	0.2	14:00–15:00	200	
		Dish washer	1500	1	19:00–20:00	1500	
		Air fryer	1000	0.2	17:00–18:00	200	
		Gaming console	300	3	19:00–22:00	900	
		Coffee machine	1000	0.3	6:00–7:00	300	
		Total		Average S = 15.8, W = 20.8 kWh		17,085	5.488159

However, the proposed deep neural network based model aims at predicting the hourly load profile of a residential building with just having the occupancy rate the input to the prediction model. In general, machine learning models require a feature (input) and a label (output) to learn and be successful in doing estimation. The randomly generated load profiles are appropriated tagged, such that it could be used to train the prediction algorithms considered in the study. The energy consumption values from the corresponding hour were then added together using Equation (1) to obtain a new version of the load profile for the different household. Hence, the hourly daily usage load profile is developed and shown in Table 9.

$$E_H = \sum_{i=0}^{n} P_i \tag{1}$$

where E_H is the energy amount in a particular hour and P_i is the power used in that hour by appliance i.

Table 9. Initial load profile created for the project (summer).

Time	Energy Consumption (Wh)	Time	Energy Consumption (Wh)
00:00	230	12:00	360
01:00	230	13:00	480
02:00	230	14:00	480
03:00	230	15:00	480
04:00	230	16:00	480
05:00	230	17:00	280
06:00	230	18:00	580
07:00	123	19:00	380
08:00	80	20:00	380
09:00	80	21:00	380
10:00	360	22:00	230
11:00	695	23:00	230

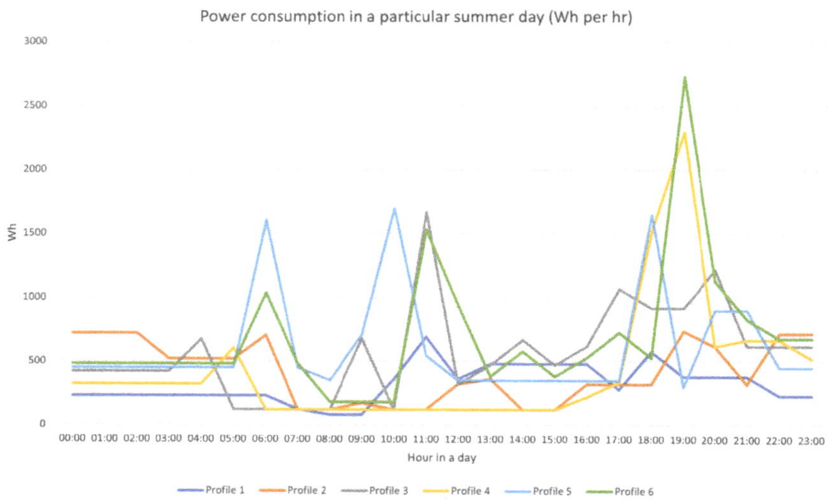

Figure 4. Visual representation of all the load profiles.

The background research indicated that the load forecasting techniques based on the occupancy rate and seasonality were niche and an area left unexplored which is considered as one of the key contribution of the proposed work. Based on which the randomised load profile generator was modelled. The load profile generator randomly generates several combinations of occupancy rate, season, and time resulting in a dataset that replicates the hourly generation pattern associated with the input feature as illustrated in Table 9. The equation used for generating the energy consumption patterns is expressed in Equation (2), where E_H is the energy amount in a particular hour; E_O is the base energy value; $random(number_{min}, number_{max})$ generates a random number between minimum and maximum values. The final load profiles can be seen in Table 10, which includes 100,000 data sets. The advantage of using the synthetic load profile generator for building more custom load profiles is that the system's randomness can be fine-tuned based on the requirement, making it a more robust and adaptable solution. This approach decreases the latency that can occur with the dataset focused on a particular context.

$$E_H = random(0.8, 1.2) * E_O + random(0, 50) \tag{2}$$

Table 10. The Final version of the load profile.

Index	Season	Number of Occupants	Hour	Energy Consumption (Wh)
1	0	1	0	230
2	0	1	1	230
3	0	1	2	230
4	0	1	3	230
5	0	1	4	230
:	:	:	:	:
99,996	0	2	2	303.92
99,997	1	1	22	272.93
99,998	0	3	18	784.27
99,999	1	1	3	177.36
100,000	1	4	7	518.85

4.2. Machine Learning Modelling

In general, machine learning models work with the dataflow of taking in input features, extracting a relationship with the input feature and the output label, and predicting the future. In our proposed model, the input features of occupancy rate, seasonality, and datetime are given to the deep neural network model as input features. The deep learning models output is essentially the estimated value of energy consumption as illustrated in Figure 5. As indicated in the previous section, the synthetic load generator was used to generate the dataset in this pattern, and then the full set of data was then fed into the proposed deep learning model. The learning requires two stages, the training stage to create the prediction model and the testing stage to verify the prediction model's prediction performance.

Furthermore, the Python programming language with Tensor flow and Keras libraries was used to develop the MLR, XGB, and shallow/basic ANN models, and the proposed deep neural network model. A different number of hyper-parameter tuning approaches were included in the shallow/simple or conventional ANN model to obtain the proposed deep neural network model. Results indicated that after the hyper-parameter tuning, the prediction accuracy of the model had improved significantly.

4.2.1. Proposed Deep Neural Network Model

A machine learning model's performance is heavily dependent on its hyper-parameters and in general, the hyper-parameters are tunable, and finding an optimised value for these parameters can directly influence the performance of the model [23]. It is essential to understand that this study focuses on optimising a shallow ANN model's hyper-parameters to obtain a more accurate and useful deep learning model. On the other hand, hyper-parameters are external parameters that are set by the operator of the network [24]. For example, there are two types of hyper-parameters: Hyper-parameters related to neural network structure (number of hidden layers, dropout, activation function, etc.) and hyper-parameters related to training algorithm (learning rate, epoch, iterations, batch size, etc.) [24]. In this study, an iterative process of fine-tuning the shallow ANN model's hyper-parameters is performed by optimising the number of hidden layers, activation function, and dropout layers to result in the proposed deep ANN model. A deep neural network, also known as a multi-layer neural network, has more hidden layers than a shallow one. Which enables the deep neural network models to learn more abstractions relationships within the input data and how the features interact with each other on a non-linear level [25].

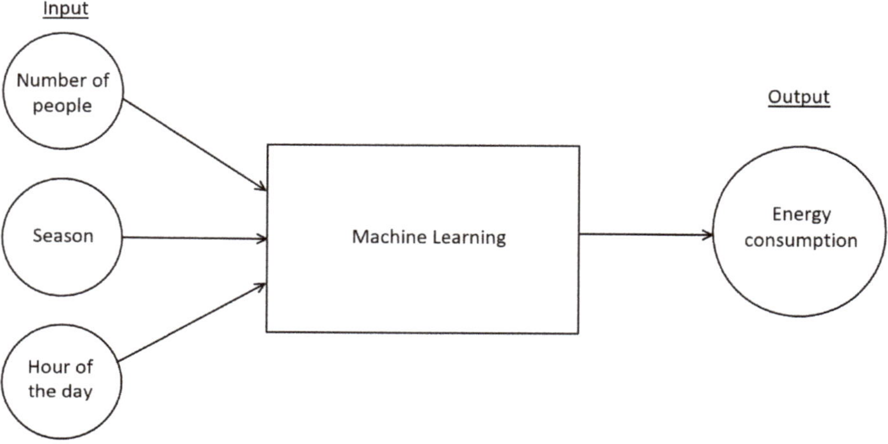

Figure 5. Data-flow of machine learning with the input and output choices.

Hidden layers are the layers of neurons in between the input layer and the output layer. Increasing the number of hidden layers increases the network model's ability. However, there is a limit to the number of hidden layers added before its effectiveness declines. Optimising this value is a challenging task in creating deep neural network models, and in this proposed model, the optimised number of hidden layers was roughly around six. Besides that, two dropout layers were also added. Dropout layers are the layers that randomly "kill" a certain percentage of neurons in every training iteration to ensure some information learned is randomly removed, reducing the risk of over-fitting the data during the training phase [24]. Having the right combination of hidden and dropout layers in ANN makes it a useful prediction model, and in this case, we are calling this developed model a deep ANN.

Additionally, the activation function is a set of rule that determines if a neuron should "fire" or not [23]. There are many types of activation functions, and each one is suited for distinctive situations. For example, a sigmoid function returns an output of "1" when the neuron's input is one or higher. Similarly, it produces a negative one when the input is below the negative one and returns the same value to the input when the input is between "−1" to "1". Rectified Linear is another activation function. This function outputs 0 when input is negative, while the output matches the input when input is positive. In this study, the sigmoid function was used for the network's hidden layers, while rectified linear function was used for the output layer.

Apart from the modification mentioned above, the number of neurons was also varied and tested to find an optimum neuron number for the effective deep ANN model. Ultimately, the study is aimed to introduce a new optimised and finely tuned deep ANN model. Figure 6 below depicts the improvements being made on the neural network to transform it from a shallow ANN model to a deep ANN model, while Figure 7 demonstrates the architecture of the optimised deep neural network.

Finally, we use root mean squared error (RMSE) and coefficient of determination (R^2) to evaluate the models created. The RMSE produces an average difference between the estimated value and the actual value. The desired RMSE value is to be as small as possible to indicate the predicting model is accurate. On the other hand, the R^2 indicates how closely the model can follow the expected estimate of energy consumption value in percentage. The R^2 value is desired to be as close to 100% as possible. The proposed deep ANN model is compared with the baseline models of XGBoost, MLR, and shallow ANN. The inferences from this evaluation are discussed in detail in the results and discussions section below.

Shallow ANN Layout

```
Layer (type)                 Output Shape              Param #
=================================================================
dense_32 (Dense)             (None, 500)               2000
_________________________________________________________________
dense_33 (Dense)             (None, 1024)              513024
_________________________________________________________________
dense_34 (Dense)             (None, 1)                 1025
=================================================================
Total params: 516,049
Trainable params: 516,049
Non-trainable params: 0
```

Deep ANN Layout

```
Layer (type)                 Output Shape              Param #
=================================================================
normalization_3 (Normalizati (None, 3)                 7
_________________________________________________________________
dense_12 (Dense)             (None, 500)               2000
_________________________________________________________________
dense_13 (Dense)             (None, 1024)              513024
_________________________________________________________________
dropout (Dropout)            (None, 1024)              0
_________________________________________________________________
dense_14 (Dense)             (None, 1024)              1049600
_________________________________________________________________
dense_15 (Dense)             (None, 1024)              1049600
_________________________________________________________________
dropout_1 (Dropout)          (None, 1024)              0
_________________________________________________________________
dense_16 (Dense)             (None, 1)                 1025
=================================================================
Total params: 2,615,256
Trainable params: 2,615,249
Non-trainable params: 7
```

Figure 6. Layout summary of our ANN designs, shallow ANN (**left**), and deep ANN (**right**).

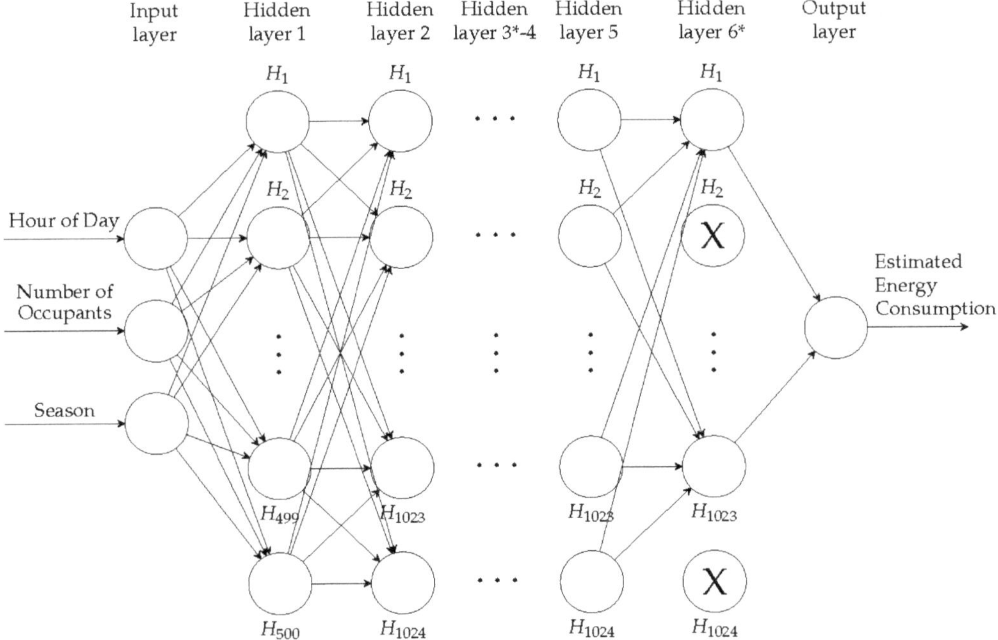

Figure 7. The proposed deep ANN layout for this study. * Layer 3 and layer 6 are the dropout layers.

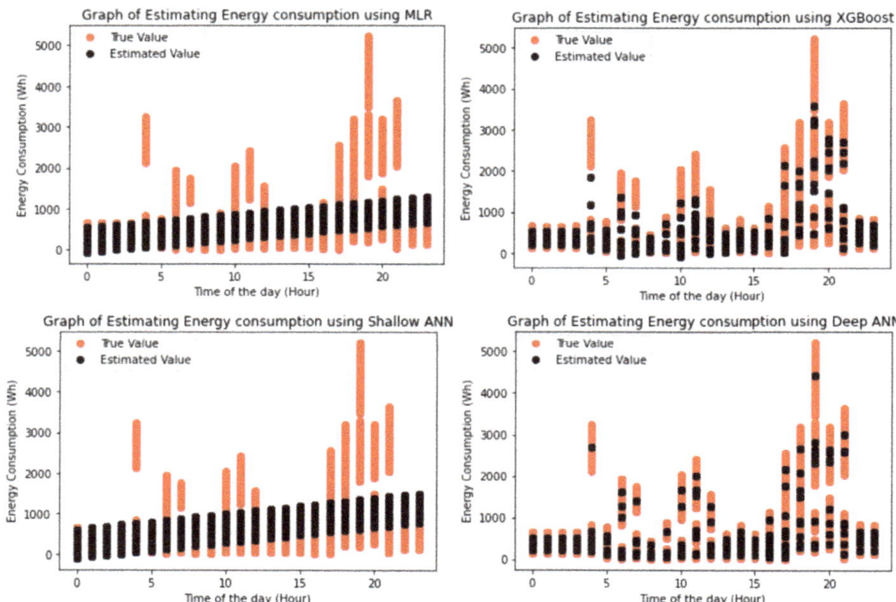

Figure 8. The graph shows the predication from each techniques: MLR (**top left**), XGB (**top right**), shallow ANN (**bottom left**), and deep ANN (**bottom right**).

5. Results and Discussion

Figure 8 illustrates the hourly error in the prediction for each techniques. The final results of the study comparing the performance of the proposed deep ANN with the other baseline approaches is shown in Table 11. It is very clear from the results that the proposed deep ANN and XGB were more accurate in prediction than the shallow ANN and MLR models. The observations reinforce the finding from the MCA analysis carried out on the literature illustrating the proposed model's acceptance. However, the result from Table 11 also emphasises that a simple shallow ANN would perform as weak as an MLR model unless it is adequately tuned. It is shown that hyperparameter tunning allowed the shallow ANN to achieve much higher accuracy, with an exceptional coefficient of determination of 97.5%.

Table 11. The result of the simulation.

Technique	RMS Error (Watts)	Coefficient of Determination (%)
Deep ANN	111.20	97.5
XGB	270.85	84.9
MLR	634.65	17.2
Shallow ANN	636.74	16.6

The MLR prediction model graph showed that the estimated values followed a linear pattern and did not adequately represent the actual values. MLR machine learning method achieved an average RMSE value of 635 W and 17% accuracy. Similarly, the shallow ANN prediction model achieved an average RMSE value of 637 W and 18% accuracy. Therefore, it is clear that a lack of hidden layers and hyperparameter tuning significantly reduces the ANN model's predicting performance. On the other hand, the XGB prediction model's estimated value was much closer to the expected values. XGBoost machine learning method achieved an average RMSE value of 271 W and 85% accuracy. Even better, the deep ANN model's estimated values were able to resemble the original data set to a greater extent. The deep ANN machine learning method achieved an average RMSE value of 111 W and 98% accuracy. Overall, this trend in the results is consistent with the inferences obtained from the MCA in Section 3.2 and the machine learning model's design objectives highlighted in Section 4.2.1.

The line graph in Figure 9 reveals a better visual representation of all models' performance. The deep ANN model achieved the highest accuracy of all the techniques examined in this study, indicated by the fact that it had the lowest RMSE curve compared to the rest. The increased accuracy is due to its ability to model the randomness in the model and deal with input noises, unlike linear regression methods, which are only suited for linear modelling. The XGB models are also susceptible to noisy data, evident from the accurate prediction results obtained. However, despite performing well in predicting energy consumption values, deep ANN took a significant longer computation time (2738 s or 45 min and 38 s) to build the model compared to 4 s and 28 s for MLR, XGB respectively. Based on the accuracy required and computational availability, the type of prediction model is chosen.

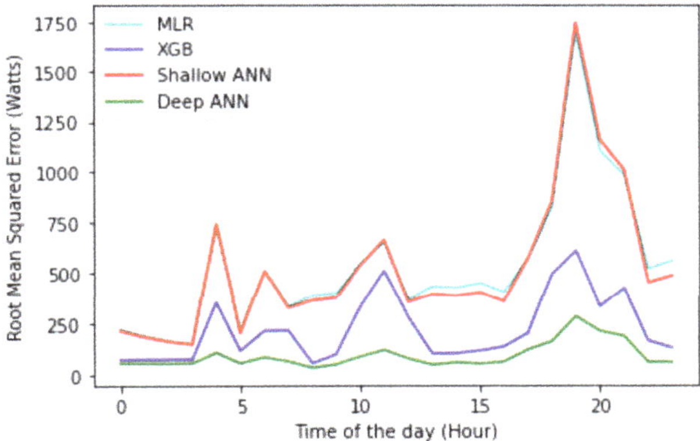

Figure 9. The graph of RMSE of estimated energy consumption value using MLR (cyan), XGBoost (blue), shallow ANN (red), and deep ANN (green).

6. Conclusions

In this study, a predication algorithm of a residential building based on the occupancy rate was investigated. The synthetic load profile generator model proposed in this study is close to realistic model were used to generate the random load profiles which were used to train the proposed state-of-the-art deep ANN model. The computation time and the accuracy of different machine learning models were then compared, and the results indicated that the proposed deep ANN model was the most appropriate for energy consumption prediction.

This study's main contribution was the novel predictive algorithm for load forecasting based on occupancy rate and the establishment of the finely tuned deep ANN model. Other findings from the research include:

- MLR was the least accurate in prediction (17.2%), but it was the fastest in computation (28 s). Since the energy consumption values do not have a close linear relationship with time but instead present significant randomness in peak consumption, it was difficult for MLR to find the best fit function, and hence, accurately predict values.
- XGBoost performs better than MLR in predicting energy consumption (84.9%) but does not handle noise well and is not suitable for time series data set. Therefore, it falls in the middle range of the ranking.

- Deep ANN performs better than shallow ANN and can take hours or days to train the data and create a prediction model. However, the model can be very accurate in prediction (97.5%) since it works well with random data set and can handle noise. It is at the top of the chart for its ability to accurately predicting energy consumption. In the cases where computation time is not a significant concern, deep ANN is highly recommended.

Further research can be performed to optimise the hyper-parameters related to the ANN model, such as learning rate, momentum, epoch, iterations, and batch size. Additionally, the novel deep neural network based forecasting algorithm proposed can be later evaluated with realistic historical load profile data in the future. Overall, the authors believe that the novel synthetic load profile generator and the finely tuned deep ANN model developed in this study would be enhancing the performance of the load profile forecasting and can be used in future with wide variety of data sets. The synthetic profile generator could be of ideal use when we are not having access to historical data where the novel MG model can assist in generating the load profiles that can be used for forecasting the hourly energy consumption profile. The prediction algorithms also provide a great opportunity to a market operator to predict their customers' energy consumption with limited inputs, to help them identify the most optimal energy generation schedule.

Author Contributions: Conceptualisation: L.H.M.T., K.H.K.C., R.L., G.S.T., M.S. and A.S.; data curation: L.H.M.T., K.H.K.C. and R.L.; formal analysis: L.H.M.T., K.H.K.C. and R.L.; investigation: G.S.T., M.S., B.H., S.M. and A.S. methodology: L.H.M.T., K.H.K.C., R.L., G.S.T., M.S., B.H., S.M. and A.S.; project administration: L.H.M.T. and G.S.T.; resources: B.H., S.M. and A.S.; software: K.H.K.C. and G.S.T.; writing—original draft: L.H.M.T., K.H.K.C., G.S.T. and R.L.; writing—review and editing: M.S., B.H., S.M. and A.S. All authors have read and agreed to the published version of the manuscript.

Funding: This research received no external funding.

Institutional Review Board Statement: Not applicable.

Informed Consent Statement: Not applicable.

Acknowledgments: Daftlogic for the typical power consumption of households published in their website, which was the base for the synthetic data generation considered in this proposed study.

Conflicts of Interest: The authors declare no conflict of interest.

Nomenclature

MLR	Multiple linear regression
XGB	Extreme gradient boost
ANN	Artificial neural network
ML	Machine learning
ANFIS	Adaptive neural fuzzy inference system
WNN	Wavelet neural network
SVM	Support vector machine
ARIMA	Auto regression integrated moving average
MCA	Multi-criteria analysis
EMS	Energy management system

References

1. Green, D.; Sonnreich, T. Centralised to De-centralised Energy—What Does it Mean for Australia. In *Infrastructure for 21st Century Australian Cities. Australian Davos Connection, Limited*; Australian Davos Connection, Limited: Melbourne, Australia, 2015; Volume 177. Available online: https://reneweconomy.com.au/centralised-decentralised-energy-mean-34072/ (accessed on 27 January 2021).
2. Lasseter, R.; Akhil, A.; Marnay, C.; Stephens, J.; Dagle, J.; Guttromsom, R.; Meliopoulous, A.S.; Yinger, R.; Eto, J. *Integration of Distributed Energy Resources. The CERTS Microgrid Concept*; Technical Report; Lawrence Berkeley National Lab. (LBNL): Berkeley, CA, USA, 2002.
3. Lantero, A. How Microgrids work. *Renew. Sustain. Energy Rev.* **2014**, *1*, 1.

4. Al-Saedi, W.; Lachowicz, S.W.; Habibi, D.; Bass, O. Power flow control in grid-connected microgrid operation using Particle Swarm Optimization under variable load conditions. *Int. J. Electr. Power Energy Syst.* **2013**, *49*, 76–85. [CrossRef]

5. Jiang, Q.; Xue, M.; Geng, G. Energy management of microgrid in grid-connected and stand-alone modes. *IEEE Trans. Power Syst.* **2013**, *28*, 3380–3389. [CrossRef]

6. Kinn, M.C. Proposed components for the design of a smart nano-grid for a domestic electrical system that operates at below 50V DC. In Proceedings of the 2011 2nd IEEE PES International Conference and Exhibition on Innovative Smart Grid Technologies, Manchester, UK, 5–7 December 2011; pp. 1–7.

7. Mateska, A.K.; Borozan, V.; Krstevski, P.; Taleski, R. Controllable load operation in microgrids using control scheme based on gossip algorithm. *Appl. Energy* **2018**, *210*, 1336–1346. [CrossRef]

8. Morsali, R.; Thirunavukkarasu, G.S.; Seyedmahmoudian, M.; Stojcevski, A.; Kowalczyk, R. A relaxed constrained decentralised demand side management system of a community-based residential microgrid with realistic appliance models. *Appl. Energy* **2020**, *277*, 115626. [CrossRef]

9. Zhang, L.; Sun, M. Research on the type of load of accessing to microgrid based on reliability. In *Advances in Computer Science Research, Proceedings of the 2015 2nd International Conference on Electrical, Computer Engineering and Electronics, Jinan, China, 29–31 May 2015*; Atlantis Press: Amsterdam, The Netherlands, 2015.

10. Fields, S. What are critical load panels? *Energysage* **2020**, *1*, 1.

11. Olivares, D.E.; Mehrizi-Sani, A.; Etemadi, A.H.; Cañizares, C.A.; Iravani, R.; Kazerani, M.; Hajimiragha, A.H.; Gomis-Bellmunt, O.; Saeedifard, M.; Palma-Behnke, R.; et al. Trends in Microgrid Control. *IEEE Trans. Smart Grid* **2014**, *5*, 2. [CrossRef]

12. Australian Government; Department of Industry, Science, Energy and Resources. Renewables. 2019. Available online: https://www.energy.gov.au/data/renewables (accessed on 18 October 2020).

13. Shi, H.; Xu, M.; Li, R. Deep learning for household load forecasting—A novel pooling deep RNN. *IEEE Trans. Smart Grid* **2017**, *9*, 5271–5280. [CrossRef]

14. Hulstaert, L. Black-box vs. white-box models. *Towards Data Sci.* **2019**, *1*, 1.

15. Sajjad, M.; Khan, Z.A.; Ullah, A.; Hussain, T.; Ullah, W.; Lee, M.Y.; Baik, S.W. A novel CNN-GRU-based hybrid approach for short-term residential load forecasting. *IEEE Access* **2020**, *8*, 143759–143768. [CrossRef]

16. Park, H.S.; Phuong, D.X.; Kumar, S. AI based injection molding process for consistent product quality. *Procedia Manuf.* **2019**, *28*, 102–106. [CrossRef]

17. Khosravani, M.R.; Nasiri, S.; Weinberg, K. Application of case-based reasoning in a fault detection system on production of drippers. *Appl. Soft Comput.* **2019**, *75*, 227–232. [CrossRef]

18. Fumo, N.; Biswas, M.R. Regression analysis for prediction of residential energy consumption. *Renew. Sustain. Energy Rev.* **2015**, *47*, 332–343. [CrossRef]

19. Tso, G.K.; Yau, K.K. Predicting electricity energy consumption: A comparison of regression analysis, decision tree and neural networks. *Energy* **2007**, *32*, 1761–1768. [CrossRef]

20. Rosenblatt, F. The perceptron: a probabilistic model for information storage and organization in the brain. *Psychol. Rev.* **1958**, *65*, 386. [CrossRef] [PubMed]

21. Kalogirou, S.A.; Bojic, M. Artificial neural networks for the prediction of the energy consumption of a passive solar building. *Energy* **2000**, *25*, 479–491. [CrossRef]

22. Fields, S. List of the Power Consumption of Typical Household Appliances. 2020. Available online: https://www.daftlogic.com/information-appliance-power-consumption.htm (accessed on 13 October 2020).

23. DeepAI. Hyperparameter. 2020. Available online: https://deepai.org/machine-learning-glossary-and-terms/hyperparameter (accessed on 27 January 2021).

24. MissingLink. The Complete Guide to Artificial Neural Networks: Concepts and Models. 2020. Available online: https://missinglink.ai/guides/neural-network-concepts/complete-guide-artificial-neural-networks/ (accessed on 27 January 2021).

25. Wagenaar, T. Deep Networks vs. Shallow Networks: Why Do We Need Depth? 2020. Available online: https://stats.stackexchange.com/q/274571) (accessed on 28 January 2021).

Article

Energym: A Building Model Library for Controller Benchmarking

Paul Scharnhorst [1,2,*], Baptiste Schubnel [1], Carlos Fernández Bandera [3], Jaume Salom [4], Paolo Taddeo [4], Max Boegli [1], Tomasz Gorecki [1], Yves Stauffer [1], Antonis Peppas [5] and Chrysa Politi [5]

[1]   Centre Suisse d'Electronique et de Microtechnique (CSEM), 2000 Neuchâtel, Switzerland; baptiste.schubnel@csem.ch (B.S.); max.boegli@csem.ch (M.B.); tomasz.gorecki@csem.ch (T.G.); Yves.STAUFFER@csem.ch (Y.S.)

[2]   Automatic Control Laboratory, EPFL, 1015 Lausanne, Switzerland

[3]   Department of Environmental Engineering, University of Navarra, 31009 Pamplona, Spain; cfbandera@unav.es

[4]   Thermal Energy and Building Performance Group, Catalonia Institute for Energy Research (IREC), 08930 Barcelona, Spain; jsalom@irec.cat (J.S.); ptaddeo@irec.cat (P.T.)

[5]   School of Mining and Metallurgical Engineering, National Technical University of Athens (NTUA), 15780 Athens, Greece; antonis.peppas@metal.ntua.gr (A.P.); chrysapol@metal.ntua.gr (C.P.)

*   Correspondence: paul.scharnhorst@csem.ch

Abstract: We introduce the Python-based open-source library Energym, a building model library to test and benchmark building controllers. The incorporated building models are presented with a brief explanation of their function, location and technical equipment. Furthermore, the library structure is described, highlighting the necessary features to provide the benchmarking and control capabilities, i.e., standardized evaluation scenarios, key performance indicators (KPIs) and forecasts of uncertain variables. We go on to characterize the evaluation scenarios for each of the models and give formal definitions of the KPIs. We describe the calibration methodologies used for constructing the models and illustrate their usage through examples.

Keywords: building control; simulation software; benchmarking

Citation: Scharnhorst, P.; Schubnel, B.; Fernández Bandera, C.; Salom, J.; Taddeo, P.; Boegli, M.; Gorecki, T.; Stauffer, Y.; Peppas, A.; Politi, C. Energym: A Building Model Library for Controller Benchmarking. *Appl. Sci.* **2021**, *11*, 3518. https://doi.org/10.3390/app11083518

Academic Editor: Elmira Jamei

Received: 16 March 2021
Accepted: 12 April 2021
Published: 14 April 2021

Publisher's Note: MDPI stays neutral with regard to jurisdictional claims in published maps and institutional affiliations.

1. Introduction

Buildings play an important role in the total energy consumption and greenhouse gas (GHG) emissions worldwide. According to [1], 36% of the global final energy is used in buildings (30% in building operation) and buildings account for 39% of GHG emissions (28% in operation, see also [2]). Alongside building renovations, smart control strategies will be key technological enablers for reducing buildings' GHG footprints and meeting the Paris Agreement goals.

The current standard for controlling Heating Ventilation and Air-Conditioning (HVAC) systems is formed by rule-based and proportional-integral (PI) controllers [3,4], but their rather simple nature, combined with possible tuning errors, can lead to sub-optimal control behavior [5]. Therefore, automated and efficient building control provides the chance to significantly reduce energy consumption and emissions. Recent research approaches cover the fields of (robust) model predictive control (MPC) (see e.g., [6–9]), adaptive or learning-based MPC (see e.g., [10]), and reinforcement learning (RL); see e.g., [11–13]. A comprehensive overview of MPC and data-driven approaches for building control can be found in [14,15].

Yet many of the aforementioned studies suffer from the non-standardized evaluation of their control performance. Some were demonstrated in simulations (e.g., [8]), others in real sites (e.g., [7]), but most of them in a single building or simulated model, and for a rather short period of time (e.g., one day and one week experiments for [8], and five day experiments for [7]). Hence, a direct comparison of the performance of the control methods is impossible. Moreover, for industrial purposes, it is desirable to create scalable controllers that can run without major modifications on different types of buildings and do not necessitate fine-tuning by hand, as often practiced in research studies.

Tools that allow to easily run simulations on different, realistic building models, provide a way to alleviate these issues. We will restrict ourselves to discussing frameworks that are able to handle the two most popular modeling languages for buildings, namely EnergyPlus [16] and Modelica [17]. Tools that are capable of including, or including these types of models are Modelicagym [18], an RL testbed for power consumption optimization [19], BOPTEST [20], and CityLearn [21]. The former two provide interfaces to the API of the popular RL benchmarking library Gym [22] for including Modelica (for the first package) or EnergyPlus models (for the second package). Their focus does not lie on providing a broad range of pre-compiled models or fixed evaluation metrics, which limits their practicality for benchmarking. BOPTEST, on the other hand, follows an approach similar to Energym, while concentrating purely on Modelica models. Of their 10 selected reference cases, two are currently available, together with four additional prototypes. The models are launched using Docker and the communication is done through a RESTful API, which differs from the Energym approach of providing pre-compiled models or a Docker container and the Gym API. Recently, the developers of BOPTEST also started to incorporate the Gym API into their framework. CityLearn tackles the problem of demand response, the coordinated control of multiple buildings to match the power demand to the provided power by the distribution grid. Aiming also at providing a benchmark for RL controllers, their focus does not lie on the assessment of single building control performance.

Contribution: We present Energym (available at https://github.com/bsl546/energym (accessed on 14 April 2021)), a Python-based library, made for standardized comparison and evaluation of controller performances, based on predefined evaluation scenarios, and inspired by the RL benchmarking library Gym. The included building models to date are based on EnergyPlus and Modelica, and are interfaced using the Functional Mock-up Interface (FMI) standard [23] through compiling them as Functional Mock-up Units (FMUs). We provide an overview of these models regarding their characteristics, location and technical equipment.

Key performance indicators (KPIs) that relate to thermal comfort, energy consumption, and emissions are defined and we lay out the standardized evaluation scenarios for each model.

Four of the 11 models already present are entirely based on real buildings (envelope and equipment were calibrated with recorded measurement data), whereas for the other models, only parts of the equipment were calibrated. The calibration methodologies are explained and supported by examples from the models.

2. Buildings Overview

Energym includes 11 simulation models to date (three Modelica models and eight EnergyPlus models). The EnergyPlus models are all updated to the current Energyplus version 9.4. An overview of the installed technical equipment and their controllability is given in Table 1. A description of each model's inputs and outputs is provided in Appendix B. The models differ in size, number of rooms, usage profile, technical equipment, controllability, and climate zone. The seven buildings that are the base for the 11 models are listed below. Four of them are available in two versions, either differing in the control (e.g., controlling thermostat setpoints vs. controlling the equipment directly) or the installed equipment. The buildings have the following characteristics.

- **Apartments**: A residential building with four stories, each being one apartment, and eight thermal zones (two per story). It is located in Spain and has a central geothermal Heat Pump (HP) providing heat to all apartments. The building envelope is fictive, based on typical Spanish construction materials, but the HP was calibrated with a real HP located in the IREC laboratory (see Section 4.2).
- **Apartments2**: This building shares its envelope with the Apartments building, but differs in the details of the technical equipment: each apartment possesses its own

air-to-water HP and its own heating storage tank. In this building, the electrical systems (solar panels, battery) were calibrated with real systems.

- **Offices**: This building consists of 25 thermal zones, 14 of which are controlled with thermostats, and is located in the Attica region of Greece. Both the envelope and the technical systems were calibrated with the corresponding test site data.
- **MixedUse**: A building with 13 thermal zones (eight controlled with thermostats) used for residential and office spaces. It is also located in the Attica region of Greece. Both the envelope and the technical systems were calibrated with the corresponding test site data.
- **Seminarcenter**: This building encapsulates 27 thermal zones, belonging to 22 rooms of which 18 are controlled with thermostats. It is located in the Region of Southern Denmark. Both the envelope and the technical systems were calibrated with the corresponding test site measurements.
- **SimpleHouse**: This building is a standard single room house with HP and sun heating effects through glazing. Two versions exist, one with a standard radiator (SimpleHouseRad) and the other one with floor heating (SimpleHouseSlab). The first order envelope model is designed based on thermal peak power and minimal outdoor temperature.
- **SwissHouse**: This building is a large-scale version of the SimpleHouse building with radiators. It has been designed with parameters (thermal peak power, outdoor temperature) from a real house. The heat pump control differs from the SimpleHouse model, see Table A6.

The calibration methodology used for the models is explained in Section 4.

Table 1. Equipment of the different models in Energym. Th: Thermostat, HP: Heat Pump, Bat: Battery, AHU: Air Handling Unit, EV: Electric Vehicle, PV: Photovoltaic. ✓: present and controllable, #: present but not controllable, ×: absent. The last column refers to the section where the calibration method used for each model is described.

Environment	Th	HP	Bat	AHU	EV	PV	Soft.	Loc.	Calib.
ApartmentsThermal-v0	✓	✓	✓	×	✓	#	E+	ESP	Section 4.2
ApartmentsGrid-v0	✓	#	✓	×	✓	#	E+	ESP	Section 4.2
Apartments2Thermal-v0	✓	✓	✓	×	✓	#	E+	ESP	Section 4.2
Apartments2Grid-v0	✓	#	✓	×	✓	#	E+	ESP	Section 4.2
OfficesThermostat-v0	✓	×	×	×	×	#	E+	GRC	Section 4.1
MixedUseFanFCU-v0	✓	×	×	✓	×	×	E+	GRC	Section 4.1
SeminarcenterThermostat-v0	✓	#	×	#	×	#	E+	DNK	Section 4.1
SeminarcenterFull-v0	✓	✓	×	✓	×	#	E+	DNK	Section 4.1
SimpleHouseRad-v0	×	✓	×	×	×	#	Mod	CHE	Section 4.3
SimpleHouseSlab-v0	×	✓	×	×	×	#	Mod	CHE	Section 4.3
SwissHouseRad-v0	×	✓	×	×	×	#	Mod	CHE	Section 4.3

3. Library Design and Functionalities

3.1. Design Features

Energym is designed to work with different controller types including rule-based controllers (RBC), MPC controllers and RL-based controllers. Hence the building environments and their interface are provided but the controller structure is not prescribed and left free to the user. Moreover, model performance evaluation is not based on fixed rewards (like in Gym) but implemented via KPIs that can be computed by the user after an evaluation run. The main features of the library are outlined below (A full documentation of the library, describing usage and installation, is available at https://bsl546.github.io/energym-pages/ (accessed on 14 April 2021)).

3.1.1. Standardized Evaluation

For each model outlined in Table 1, a physical objective to be reached is predefined. This objective might be, e.g., the minimization of the CO_2 emissions related to the building operation. The controllers also have to satisfy thermal constraints to guarantee occupant comfort. These two quantities—objective and constraints—are tracked with the implemented KPIs; see Section 3.3, Table 2. For each building, the evaluation phase with the predefined KPIs is run over a definite period of time and under predefined weather conditions.

Table 2. Fixed evaluation scenarios for the simulation models.

Model	Simulation Period	Temperature Constraints (°C)	Objective KPI
ApartmentsThermal-v0	January–April	19–24	Grid exchange
ApartmentsGrid-v0	Entire year	19–24	Grid exchange
Apartments2Thermal-v0	January–April	19–24	Grid exchange
Apartments2Grid-v0	Entire year	19–24	Grid exchange
OfficesThermostat-v0	Entire year	19–24	Power demand
MixedUseFanFCU-v0	Entire year	19–24	Power demand
SeminarcenterThermostat-v0	January–May	21–24	CO_2 emissions
SeminarcenterFull-v0	January–May	21–24	CO_2 emissions
SimpleHouseRad-v0	January–April	19–24	Power demand
SimpleHouseSlab-v0	January–April	19–24	Power demand
SwissHouseRad-v0	January–April	19–24	Power demand

3.1.2. Wrappers

Wrappers are implemented to cope with different controller needs. In particular, wrappers are provided to scale inputs and/or outputs between values in a min-max fashion. The scaling can be beneficial for optimization-based controllers like MPC, due to the used model and solver structure. For RL-controllers, an RL-wrapper is provided to change the outputs of the `step` method and provide exactly the same outputs as in the Gym library, i.e., `outputs, reward, done, info`. One slight change with respect to Gym, however, is that the reward design is left free to the user and must be specified at wrapper initialization. This design choice was made for users to be free in the reward design phase, the main objective of any controller being to minimize the predefined KPIs. Similarly, for controller speed-up (in particular for MPC), a downsampling wrapper is provided to optimize computation time, making it possible to solve the problem less frequently than what the standard `step` method would impose.

3.1.3. Forecasting Capabilities

For designing controllers such as MPC, it is important to have descriptions of external disturbances. For this, we provide weather forecasts (including irradiance and temperatures), optionally given by the exact values in the used weather files or by stochastic variations of those. Furthermore, we provide forecasts that are highly relevant for certain models: EV usage schedules for the Apartments and Apartments2 buildings, and electricity mix forecasts for the Seminarcenter. Random seeds to generate the forecasts are fixed in evaluation mode to ensure reproducibility of the results.

3.2. Usage and Code Example
Basic Structure and Usage

After importing Energym, a model can be created by calling the `make` method and specifying the name of the model and other optional parameters, i.e., the starting day of the simulation, the number of simulated days, the used weather file, and the used KPIs, all of which use default values if not specified upon initialization. The interaction with the model, i.e., passing control inputs and receiving outputs, is done with the `step` method. Control inputs are Python dictionaries, with the setpoint name as key and input as value (possibly a list with multiple entries for multiple consecutive inputs), e.g., `{"Z01_T_Thermostat_sp":[21]}` (or `{"Z01_T_Thermostat_sp":[21, 22, 21]}`). Outputs are also defined as dictionaries using the predefined output names as keys. The main inputs and outputs for each model are given in Appendix B. A full list is available in the online documentation. The `Wrapper` class is implemented to provide input-output wrapper functionalities. Weather and stochastic disturbances forecasts are available with the `get_forecast` method.

For the tracking of the KPIs, a `KPI` object is initialized for each model, it automatically records the necessary data. Calling the method `get_kpi` returns the evaluation for a specified time interval (by default all the completed steps) as a dictionary. More details on handling the KPIs and the default ones are discussed in Section 3.3.

A simple usage example is given in Appendix A.1.

3.3. Performance Evaluation

A pre-compiled FMU is provided for each building model and can be used with different weather files. This allows the users to train their controllers (i.e., RL agents or models for MPC) with different weather files, while the weather file for final evaluation is fixed. These fixed weather conditions on a predefined period of time ensure comparability of the control performances via the implemented KPIs. The characteristics of these fixed evaluation scenarios are displayed in Table 2. The defined KPIs fall into the categories of thermal comfort (related to temperature constraints) and objective KPI (related to the objective to minimize).

KPI Definition

For the thermal comfort, a range of acceptable temperatures is defined. The tracked KPIs are the average deviation from the target temperatures for each controlled thermal zone and the total number of range violations. Let the desired temperature range be defined by the interval $I = [a, b]$. Then the average deviation $d(T, I)$ for temperature measurements $T = \{t_i : i = 1, \ldots, N\}$ is defined as

$$d(T, I) := \frac{1}{N} \sum_{i=1}^{N} \|t_i\|_I \tag{1}$$

where $\|t\|_I = \begin{cases} a - t, & \text{if } t < a \\ 0, & \text{if } t \in I \\ t - b, & \text{if } t > b \end{cases}$.

The number of total violations $v(T, I)$ is defined as

$$v(T, I) := \sum_{i=1}^{N} \delta(t_i, I) \tag{2}$$

where $\delta(t, I) = \begin{cases} 0, & \text{if } t \in I \\ 1, & \text{if } t \notin I \end{cases}$.

The average energy exchanged with the grid is tracked for the models based on the Apartments and Apartments2 buildings. Let $E_{prod} = \{e_{prod,i} : i = 1, \ldots, N\}$ be the set of N consecutive measurements of produced energy and $E_{con} = \{e_{con,i} : i = 1, \ldots, N\}$ of consumed energy. Then the average energy exchange $e(E_{prod}, E_{con})$ is defined as

$$e(E_{prod}, E_{con}) := \frac{1}{N} \sum_{i=1}^{N} |e_{prod,i} - e_{con,i}|. \tag{3}$$

In the evaluation scenario, the goal is to minimize this quantity and therefore maximize the self-consumption of produced energy.

The objective for the Offices, MixedUse, SimpleHouse and SwissHouse buildings is to minimize their power consumption. Let the mean power demand for N simulation steps be given by $D = \{d_i : i = 1, \ldots, N\}$. The minimization objective is again given by averaging over the measurements, so the average power demand $p(D)$ is defined as

$$p(D) := \frac{1}{N} \sum_{i=1}^{N} d_i. \tag{4}$$

The environments based on the Seminarcenter building track the CO_2 emissions for the installed gas boiler and the varying electricity mix. A minimization of this emission is the focus of their evaluation scenario. Let the emission values be given by $C = \{c_i : i = 1, \ldots, N\}$. The computed KPI for those measurements is the average emission $g(C)$ defined as

$$g(C) := \frac{1}{N} \sum_{i=1}^{N} c_i. \tag{5}$$

Instead of using predefined KPIs, it is also possible to define custom KPIs. An example of this is given in Appendix A.2.

4. Calibration Methodology

Distinct methodologies (see Table 1) for calibration have been used, depending on whether the entire building or just the technical systems were calibrated with real data sets. After calibration and validation with standard metrics (see e.g., Section 4.1.2), the model responses to control actions were further tested independently by team members to ensure that physical expectations were met (setpoint responses, energy consumption patterns by system activation, etc.).

4.1. Building Calibration

The Offices, MixedUse and Seminarcenter buildings have been calibrated using the three-step methodology presented in [24]. A short overview of the method is explained in the following.

4.1.1. Method

In the first step, data are collected from the test sites at a 15 min sampling rate. Collected features include weather parameters (outside temperature, ground temperature, relative humidity, irradiance, atmospheric pressure, wind speed), indoor climate (room temperatures and relative humidity), as well as technical equipment parameters (water temperature and flow, on/off status) and electric consumption disaggregated by sources. Standard data pre-processing techniques are applied to the collected data to improve their usability, namely: gap reconstruction (via interpolation), removal of sensor malfunction periods, and on/off status reconstruction for technical systems for which this signal was not made directly available.

In a second step, the buildings are modelled and their envelope calibrated using the collected data. It uses free oscillation data, i.e., periods where the HVAC equipment is off to eliminate HVAC interference. As described in depth in [24], envelope calibration is realized through a parametric analysis, a sensitivity analysis and a genetic algorithm simulation (using the NSGA-II genetic algorithm; see [25]) guided by an appropriate objective function (based on normalized root mean square error (NRMSE), determination coefficient, and normalized mean bias error (NMBE)).

In a third step, HVAC and technical equipment are introduced and configured to supply the building demand. A detailed HVAC model is added to the previously calibrated envelope model and simulated in EnergyPlus. Known HVAC equipment parameters are set to technical specification values, different performance curves are determined for each of the components, while unknown parameters are set either based on technical information of similar equipment, or calculated based on test site data. This detailed model undergoes a new calibration process similar to the one used for the envelope, i.e., a parametric/sensitivity analysis followed by a genetic algorithm simulation guided by a new objective function. This calibration is performed until the simulation model uncertainty indices are acceptable within the expected KPIs; see Section 4.1.2.

4.1.2. Model Evaluation

Three metrics were used to assess the quality of building models: the NMBE, the coefficient of variation of the RMSE (CVRMSE) and the coefficient of determination R^2;

see e.g., [26]. Model acceptance is based on the threshold values recommended by the American Society of Heating, Refrigerating and Air-Conditioning Engineers (ASHRAE) and the International Performance Measurement and Verification Protocol (IPMVP); see e.g., [26].

4.1.3. Example: The MixedUse Building

Calibration of the envelope parameters for the MixedUse building has been performed with free oscillation data as described in Section 4.1.1. The HVAC system of the MixedUse building is made of two independent main technical systems: a Variable Refrigerant Volume Unit (VRV) and an air-to-water HP. The initial performance curves of these two systems have been fitted with linear estimations to reproduce the suppliers' technical documentation with the performance equations of the corresponding EnergyPlus objects. For the MixedUse VRV system a total of 14 different curves were required, from cooling and heating capacity for low/high outdoor conditions including its boundary curves to piping length correction and defrosting; see Figures A5 and A6 in Appendix C.1.

Key parameters were then estimated in the next step (nominal power, design airflow, design supply temperature, efficiency, ...) via the optimization process with the NSGA-II genetic algorithm in order to find the combination of parameter values that results in the best fit of energy consumption while maintaining the indoor climate of the building. The results for the VRV system are displayed in Table A7 and the results for the HP system in Table A8 in Appendix C.1.

Finally, the evaluation period took place during Summer 2020, between the months of June and August (i.e., on data not used for identification). Results are displayed in Table 3. It should be noted that during the evaluation period the HP underwent a series of malfunctions and had to be repaired. This is why evaluation results for the HP are not presented here.

Table 3. Results of Variable Refrigerant Volume (VRV) and Air Handling Unit (AHU) system evaluation for cooling period, June to August 2020.

Index	ASHRAE	Cooling Enrgy	Average Indoor	Cooling Supply	TZ-05 Indoor
(Hourly)		Consumption (VRV)	Temperature (Main Building)	Air (AHU)	Temperature (Atrium)
Date	—	June–August 2020	June–August 2020	May–July 2020	May–July 2020
NMBE (%)	within 10%	6.40%	0.33%	1.43%	0.42%
CV(RMSE) (%)	≤30%	29.26%	1.16%	3.62%	1.69%
R2 (%)	≥75%	75.23%	92.73%	85.57%	86.98%

4.2. Heat Pump Calibration

For the Apartments building, heating is covered by means of a centralized water-to-water geothermal HP system that provides hot water for the indoor fan coil units and the Domestic Hot Water (DHW) tanks. This HP model has been calibrated using real data from a test bench facility installed at the IREC laboratory. The method used is based on the work presented in [27] for HP identification and the water-to-water HP model developed in [28]. Equations (6) and (7) from [28] displayed below represent the fitting of the heating thermal power and of the electric power consumed by the HP. The experimental data used to fit the equations have been obtained by operating the HP at full load in heating mode (control of return temperature to the condenser of the HP).

$$\frac{Q_h}{Q_{h_{ref}}} = D1 + D2\left[\frac{T_{L_{in}}}{T_{ref}}\right] + D3\left[\frac{T_{S_{in}}}{T_{ref}}\right] + D4\left[\frac{\dot{V}_L}{\dot{V}_{L_{ref}}}\right] + D5\left[\frac{\dot{V}_S}{\dot{V}_{S_{ref}}}\right] \tag{6}$$

$$\frac{P_h}{P_{h_{ref}}} = E1 + E2\left[\frac{T_{L_{in}}}{T_{ref}}\right] + E3\left[\frac{T_{S_{in}}}{T_{ref}}\right] + E4\left[\frac{\dot{V}_L}{\dot{V}_{L_{ref}}}\right] + E5\left[\frac{\dot{V}_S}{\dot{V}_{S_{ref}}}\right] \tag{7}$$

where:

- $D1$–$D5$, $E1$–$E5$: Fitting coefficients for heating mode
- $T_{ref} = 283.15$: Reference temperature, [K]
- $T_{L_{in}}$: Temperature of water entering the load side, [K]
- $T_{S_{in}}$: Temperature of water entering the source side, [K]
- $\dot{V}_L$: Load side volumetric flow rate, [m^3/s]
- $\dot{V}_{L_{ref}}$: Reference load side volumetric flow rate, [m^3/s]
- $\dot{V}_S$: Source side volumetric flow rate, [m^3/s]
- $\dot{V}_{S_{ref}}$: Reference source side volumetric flow rate, [m^3/s]
- P_h: Power consumption [kW]
- $P_{h_{ref}} = 3.89$: Rated heating power consumption [kW]
- Q_h: Condenser heating rate [kW]
- $Q_{h_{ref}} = 42.49$: Rated condenser heating rate [kW]

A constant volumetric flow rate was used in the experiment as the pump was operating with a constant flow, hence $\dot{V}_L = \dot{V}_{L_{ref}}$ and $\dot{V}_S = \dot{V}_{S_{ref}}$. Coefficients were fitted to the data using ordinary least squares. The Q_h calculation residuals range from 0.01–3.94% of the Q_h value (max. deviation of 1.45 kW for a nominal consumption of 36.8 kW heating power) and are displayed in Figure 1a. The corresponding R^2 value for the heating power fitting is 0.985. P_h residuals range from 0.02–4.78% of the electric power consumption value (max. deviation of 0.29 kW for an electrical power consumption of 6.24 kW). The R^2 value for the electric power fitting is 0.988. The residuals are depicted in Figure 1b.

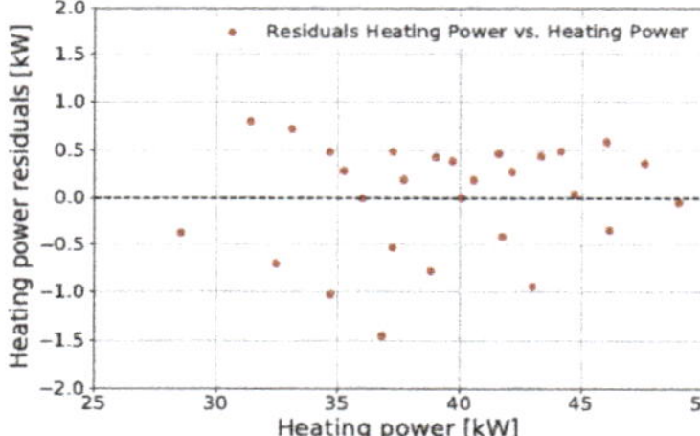

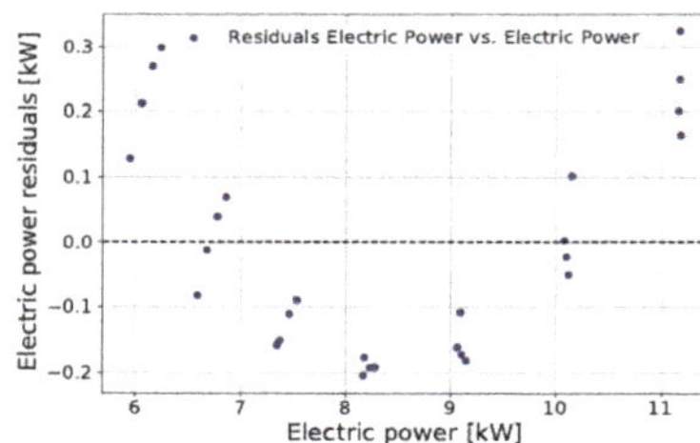

(**a**) Heating power residuals.   (**b**) Electrical power consumption residuals.

Figure 1. Residuals from the Heat Pump (HP) calibration.

4.3. Modelica Models

Modelica models are developed with components from the LBNL Modelica Buildings Library [29]. While one of the strengths of EnergyPlus is the ease at which large and realistic envelope models can be built, Modelica models with large and complex envelopes are harder to design: The strength of Modelica is the realism and flexibility at disposal for modeling and controlling technical systems like HPs, storage tanks, and AHUs. This is the reason why the currently included models come with very simple envelopes, but complex and realistic technical systems, the other case being covered by the EnergyPlus models at disposal. Future Modelica models with more complex envelopes are in preparation and will be integrated to the library. The authors also do not exclude incorporating models using Modelica for the control systems and Energyplus envelopes.

The envelope model used for SimpleHouse and SwissHouse is a simple first order model calibrated with the thermal peak power, minimum outdoor temperatures and the building free oscillation time constant. The indoor temperature is averaged over the house geometry and modeled by a scalar $T(t)$ obeying Equation (8).

$$C\frac{dT}{dt}(t) = Q_{\text{flow}}(t) - G(T(t) - T_{out}(t)) \tag{8}$$

Assuming equilibrium at very cold temperatures, the thermal conductance G [W/K] is deduced by setting the left-hand side equal to zero in these conditions. G [W/K] is then inferred to be equal to the ratio of the thermal peak power over the indoor-outdoor temperature difference for the four coldest consecutive days on the last 20 years (see SIA norm CT 2028:2010 [30]). Knowing the thermal conductance G and inferring the time constant τ of the envelope from the available building data, we derive the heat capacity C [J/K] = $G\tau$.

For SwissHouse models, these G and C correspond to an overall U-value of 0.5 [W/m^{2}K] and to a heat capacity per surface unit of 0.2 [MJ/m^{2}K], resulting in a time constant τ of >4 days, for this type of low energy building with a heat demand of 15 [W/m^2]. For air-to-water HP, the typical seasonal coefficient of performance (COP) is 3, which gives a nominal electric power of 5 [W/m^2].

5. Conclusions and Future Directions

The library Energym presented in this paper aims at providing building models and standardized evaluation scenarios and metrics to develop, test, and benchmark controllers. With diverse equipment configurations calibrated with real measurement data, Energym has been designed to ease the development and deployment of "swiss knife" data-driven controllers for buildings. The used calibration methods and results have been outlined.

Two main axes of research are foreseen for future works: The extension of the library itself and the development of data-driven control methods tested on the library.

For the former, Energym is aimed at growing by gaining new building models. We are currently developing new models (both Energyplus and Modelica models) that will be incorporated into the library in the near future. Moreover, through releasing Energym as open-source, we encourage model contributions from the building simulation community (To add a new model, please contact the authors.) and the authors welcome such efforts.

For the latter, MPC and RL-based control strategies will be extensively tested on many buildings of the library to showcase its benchmarking capabilities. Furthermore, since running models in parallel is possible with Energym, we aim to investigate scenarios with multiple models in a district setting and related control problems. Finally, an additional goal of Energym is to increase engagement within the Machine Learning community, in particular the RL community, to problems related to reducing energy consumption and CO_2 emissions.

Author Contributions: Conceptualization, P.S. and B.S.; methodology, C.F.B., M.B., T.G. and P.T.; software, P.S., B.S., C.F.B., M.B., T.G. and P.T.; validation, P.S., B.S., C.F.B., M.B., T.G. and P.T.; resources, A.P. and C.P.; data curation, C.F.B., P.T., A.P. and C.P.; writing—original draft preparation, P.S., B.S., C.F.B. and P.T.; writing—review and editing, P.S., B.S., C.F.B., M.B., T.G., P.T., J.S., A.P., C.P. and Y.S.; visualization, P.S., B.S., C.F.B. and P.T.; supervision, B.S., C.F.B., J.S. and Y.S.; project administration, J.S. and Y.S. All authors have read and agreed to the published version of the manuscript.

Funding: This project has received funding from the European Union's Horizon 2020 research and innovation program under grant agreement n°731211, project SABINA.

Data Availability Statement: Not applicable.

Acknowledgments: We thank Stephan Dasen for his strong involvement in the models testing phase, as well as Insero for the work done on the Danish site.

Conflicts of Interest: The authors declare no conflict of interest.

Appendix A. Code Examples

Appendix A.1. Basic Usage Example

A simple example of the usage of the library is given below. It demonstrates the interaction with the simulation model for 100 timesteps, assuming a function `get_input()` has been implemented, that computes the control input for the current measured state of the model and a forecast for the next 10 timesteps. The chosen parameters are arbitrary and just fulfill demonstrative purposes.

```python
import energym

env = energym.make("Apartments2Grid-v0")
out = env.get_output()
for i in range(100):
forecast = env.get_forecast(forecast_length=10)
inp = get_input(out, forecast)
out = env.step(inp)
kpis = env.get_kpi()
env.close()
```

Appendix A.2. KPI Example

Default KPIs are defined for each model, but the user can also define custom KPIs to be tracked. This is done by specifying a Python dictionary containing the information of the variables of interest and KPI computation method. An example dictionary for the KPIs looks as follows.

```python
kpi_dict = {"kpi1": {"name": "Fa_Pw_All", "type": "avg"},
"kpi2": {
"name": "Z01_T",
"type": "tot_viol",
"target": [19,24],
}
}
```

For more information on the KPI implementation, we refer to the documentation.

Appendix B. Building Descriptions

In this part, we give a short description of the buildings, and the inputs and the outputs of the simulation models that are related to the KPIs (other outputs not entering in KPIs calculation, like flow rate and flow temperature, are not listed). The common output variables for all EnergyPlus based models are given in Table A1. Complete input/output references and in-depth explanations of the buildings can be found in online documentation. The bounds given in the tables are not used to cut-off values (unless the specific cut-off wrapper is used), but are used by default by the inputs/outputs scaling wrappers to scale the signals in values close to/within the [0,1] interval.

Table A1. Common outputs for the EnergyPlus based models.

Variable Name	Description	Bounds	Units
Ext_T	Current outdoor temperature	$[-25,40]$	°C
Ext_RH	Current outdoor relative humidity	$[0,100]$	%RH
Ext_Irr	Current direct normal irradiance	$[0,1000]$	$W{\cdot}m^{-2}$

Appendix B.1. Apartments and Apartments2 Buildings

The envelope is the same for both Apartments and Apartments2 buildings; see Figure A1. The envelope is made of building elements used in the periods from 1991 to 2007 in Spain. The building model consists of four identical apartments split in two thermal zones (Figure A1). The active surface area of the PV panels is $58m^2$ with an inclination of 40° and south oriented. The PV EnergyPlus component has a rated electric power output of 10.75 kW and the inverter efficiency is 0.95. In addition, occupancy, appliances and lighting consumptions follow stochastic profiles that differentiate each dwelling behavior, resulting in different energy demands. The DHW profiles are based on the European standard (EN16147, 2011).

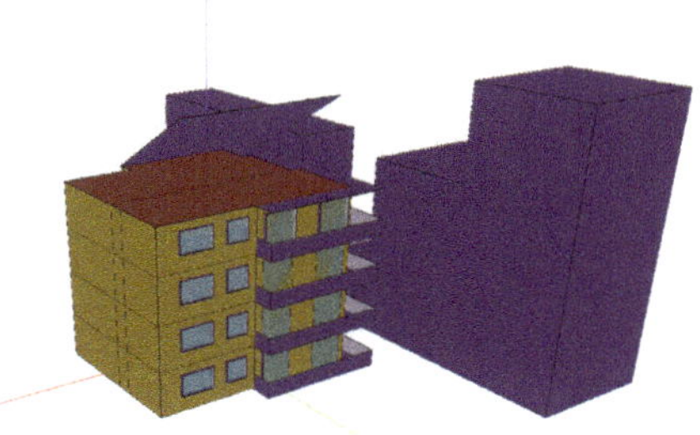

Figure A1. Envelope visualization for the Apartments and Apartments2 buildings.

The difference between Apartments and Apartments2 lies in their thermal systems. Apartments has a central geothermal HP, directly connected to hot water tanks (1 per Apartment) used only for DHW consumption, and to a heating loop providing heat to the entire building. Apartments2 does not have this central heating system, but possesses four storage tanks (supplying heating and DHW to each apartment), each being alimented by a dedicated air-to-water HP.

Both buildings possess a stationary battery with a capacity of 10 kWh, maximum power for charging and discharging of 4 kW. In apartments, there is one electric vehicle with a capacity of 20 kWh and a maximum power for charging of 3.7 kW. For Apartments2, two EVs with the same characteristics are present. Usage schedules are stochastic and forecasts are provided via the forecast API.

The evaluation weather file used for Apartments and Apartments2 is given by the identifier `ESP_CT_Barcelona_ElPratAP1` and should not be used in the training process. Control inputs and the most relevant outputs for the Apartments and Apartments2 models are listed in Table A2.

Table A2. Inputs and outputs for the models ApartmentsThermal-v0 (1), ApartmentsGrid-v0 (2), Apartments2Thermal-v0 (3) and Apartments2Grid-v0 (4).

Variable Name	Description	Bounds	Units	Model
Inputs				
P1_T_Thermostat_sp ... P4_T_Thermostat_sp	Temperature setpoint per appartment	[16,26]	°C	1/2/3/4
Bd_T_HP_sp	Heat pump supply temperature setpoint	[35,55]	°C	1/2
P1_T_Tank_sp ... P4_T_Tank_sp	Bottom water tank temperature setpoint	[30,70]	°C	1/2
Bd_Pw_Bat_sp	Battery charging/discharging setpoint	[−1,1]	-	1/2/3/4
Bd_Ch_EVBat_sp	EV battery charging setpoint	[0,1]	-	1/2
Bd_Ch_EV1Bat_sp	EV battery charging setpoint	[0,1]	-	3/4
Bd_Ch_EV2Bat_sp	EV battery charging setpoint	[0,1]	-	3/4
HVAC_onoff_HP_sp	Heat pump on/off setpoint	{0,1}	-	1
P1_onoff_HP_sp ... P4_onoff_HP_sp	Heat pump on/off setpoint	{0,1}	-	3
Outputs				
Fa_E_self	Energy exchanged with grid for timestep	[−2000,2000]	Wh	1/2/3/4
Z01_T ... Z08_T	Current zone temperature	[10,40]	°C	1/2/3/4

Appendix B.2. Offices Building

The Offices building is located in Greece and includes 25 conditioned rooms with a total area of 643.73m^2 (see Figure A2). Of those 25 rooms, 14 are controllable with thermostats (2 storage rooms, 2 lobbies, 4 seminar rooms, 1 meeting room, and 5 offices). Water-to-air fan coil units are used to condition the spaces, where either water heating is provided by an oil boiler or water cooling by an electrical air-to-water chiller.

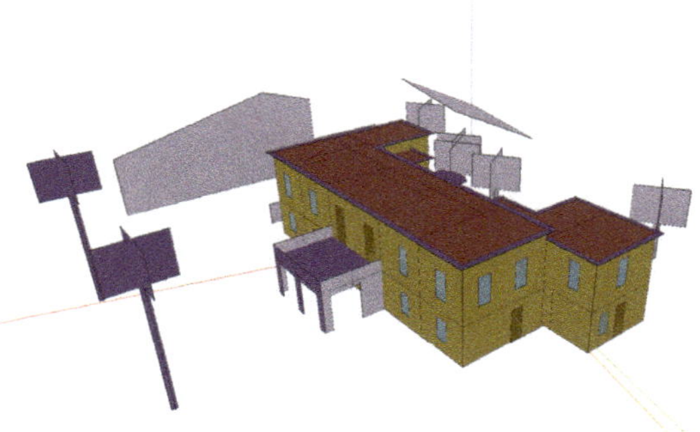

Figure A2. Envelope of the Offices building.

The evaluation weather file for the Offices building is given by the identifier GRC_TC_Lamia1 and should not be used in the training process. The inputs and some outputs are described in Table A3.

Table A3. Inputs and outputs for the model OfficesThermostat-v0.

Variable Name	Description	Bounds	Units
Inputs			
Z01_T_Thermostat_sp ... Z07_T_Thermostat_sp Z15_T_Thermostat_sp ... Z20_T_Thermostat_sp Z25_T_Thermostat_sp	Zone temperature setpoint	[16,26]	°C
Bd_Cooling_onoff_sp	Chiller on/off	{0,1}	-
Bd_Heating_onoff_sp	Boiler on/off	{0,1}	-
Outputs			
Fa_Pw_All	Current power demand of whole facility	[0,10,000]	W
Fa_Pw_PV	Current produced power	[0,2000]	W
Z01_T ... Z07_T Z15_T ... Z20_T Z25_T	Current zone temperature	[10,40]	°C

Appendix B.3. MixedUse Building

The MixedUse building is a 566.38m^2 building located in Greece with 13 thermal zones, of which eight are controllable with thermostats (see Figure A3). The HVAC system installed consists of two AHUs, one dedicated exclusively to thermal zones 5, 6 and 7 and a second one serving to the remaining thermal zones.

The first system dedicated to TZ-5, 6 and 7, is composed of an air loop, an AHU that includes water coils and two supply water loops: one with a Heat Pump Water Heater (HPWH) and the other with a chiller for cooling.

The second system serving the entire facility, consists of an air loop with an AHU that has direct expansion "DX" coils. In addition, the zones that are affected under this system have variable refrigerant flow (VRF) terminal units as part of the air-conditioning system.

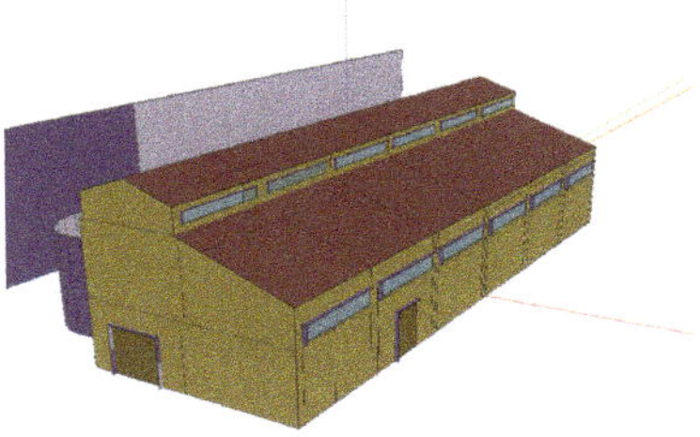

Figure A3. Envelope of the MixedUse building.

The evaluation weather file for the MixedUse building is given by the identifier GRC_TC_Lamia1 and should not be used in the training process. The control inputs and KPI related outputs are displayed in Table A4.

Table A4. Inputs and outputs for the model MixedUseFanFCU-v0.

Variable Name	Description	Bounds	Units
Inputs			
Z02_T_Thermostat_sp . . . Z05_T_Thermostat_sp Z08_T_Thermostat_sp . . . Z11_T_Thermostat_sp	Zone temperature setpoint	[16,26]	°C
Bd_T_AHU1_sp Bd_T_AHU2_sp	AHU temperature setpoint	[10,30]	°C
Bd_Fl_AHU1_sp Bd_Fl_AHU2_sp	AHU flow rate setpoint	[0,1]	-
Outputs			
Fa_Pw_All	Current power demand of whole facility	[0,50,000]	W
Z02_T . . . Z05_T Z08_T . . . Z11_T	Current zone temperature	[10,40]	°C

Appendix B.4. Seminarcenter Building

The Seminarcenter building is a one story building located in Denmark and includes 22 conditioned rooms on 1278.94m^2 (see Figure A4). Five of the 22 rooms are divided into two thermal zones and 18 rooms are controllable with thermostats.

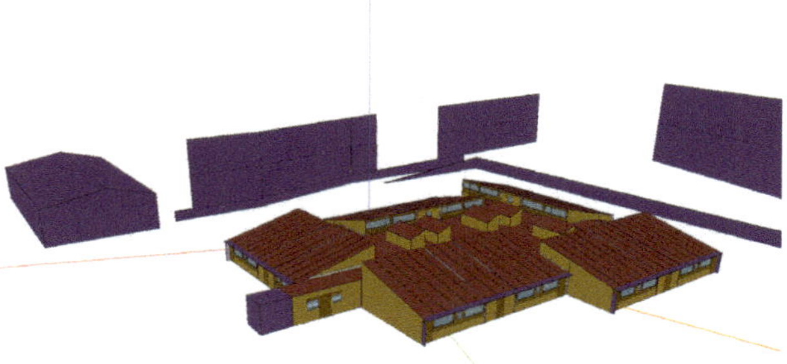

Figure A4. Envelope of the Seminarcenter building.

Heating of the rooms is provided by water convectors with hot water from a buffer tank. For the buffer tank and the DHW, air-to-water HPs are used to supply the heating demand and an additional gas boiler is available in case the HPs can not provide enough heating.

The evaluation weather file for the Seminarcenter buildings is given by the identifier DNK_MJ_Horsens2 and should not be used in the training process. The control inputs to both simulation models and some outputs are described in Table A5.

Table A5. Inputs and outputs for the models SeminarcenterThermostat-v0 (1) and SeminarcenterFull-v0 (2).

Variable Name	Description	Bounds	Units	Model
Inputs				
Z01_T_Thermostat_sp . . . Z06_T_Thermostat_sp Z08_T_Thermostat_sp . . . Z11_T_Thermostat_sp Z13_T_Thermostat_sp . . . Z15_T_Thermostat_sp Z18_T_Thermostat_sp . . . Z22_T_Thermostat_sp	Zone temperature setpoint	[16,26]	$^{\circ}$C	1/2
Bd_onoff_HP1_sp . . . Bd_onoff_HP4_sp	Heat pump on/off setpoint	{0,1}	-	2
Bd_T_HP1_sp . . . Bd_T_HP4_sp	Heat pump temperature setpoint	[30,65]	$^{\circ}$C	2
Bd_T_AHU_coil_sp	AHU water coil temperature setpoint	[15,40]	$^{\circ}$C	2
Bd_T_buffer_sp	Buffer tank temperature setpoint	[15,70]	$^{\circ}$C	2
Bd_T_mixer_sp	HPs water loop supply temperature setpoint	[20,60]	$^{\circ}$C	2
Bd_T_HVAC_sp	AHU air supply temperature setpoint	[10,26]	$^{\circ}$C	2
Outputs				
Bd_CO2	Timestep equivalent CO_2 emission mass	[0,10]	kg	1/2
Fa_Pw_All	Current power demand of whole facility	[0,100,000]	W	1/2
Z01_T . . . Z06_T Z08_T . . . Z11_T Z13_T . . . Z15_T Z18_T . . . Z22_T	Current zone temperature	[10,40]	$^{\circ}$C	1/2

Appendix B.5. SimpleHouse and SwissHouse

SimpleHouse and SwissHouse represent two residential one-family houses. The entire house is modeled with a single thermal zone in both cases. A Carnot heat pump is connected to the room model and provides heat via radiator for the rad models (SimpleHouseRad-v0 and SwissHouseRad-v0), or via floor heating for the slab model.

The evaluation weather file for the SimpleHouse and Swisshoue buildings is given by the identifier CH_ZH_Maur and should not be used in the training process. The control inputs to both simulation models and some outputs are described in Table A6.

Table A6. Inputs and outputs for the models SimpleHouseRad-v0 and SimpleHouseSlab-v0 (1), and SwissHouseRad-v0 (2).

Variable Name	Description	Bounds	Units	Model
	Inputs			
u	Heat pump normalized power	[0,1]	-	1
heaSup.f	Heat pump normalized supply flow	[0,1]	-	2
heaSup.T	Heat pump supply temperature	[293.15,353.15]	K	2
	Outputs			
TOut.T	Outside Temperature	[253.15,343.15]	K	1/2
temRoo.T	Room Temperature	[263.15,343.15]	K	1/2
heaPum.P	Heat pump power	[0,30]	kW	1/2
temRet.T	Heat pump return temperature	[273.15,353.15]	K	1/2
temSup.T	Heat pump supply temperature	[273.15,353.15]	K	1/2

Appendix C. Calibration Plots

Appendix C.1. MixedUse HVAC Performance Curves

For cooling mode, the obtained VRV capacity curves have a CV(RMSE) of 0.05%(Low), 0.10%(High) and 0.09%(Boundary) with an R2 above 99% for the three cases. While for its electric input curves it has a CV(RMSE) of 3.59%(Low), 3.06%(High) and 0.07%(Boundary) with an R2 of 91%, 96% and 99% respectively. As for heating mode, the results for the equipment capacity curves have a CV(RMSE) of 0.35%(Low), 1.07%(High) and 0.01%(Boundary) with an R2 above 99% for the three cases. The electric input curves CV(RMSE) range from 0.01%(Boundary), 0.41%(Low) to 0.75%(High), with an R2 above 99% for the three cases.

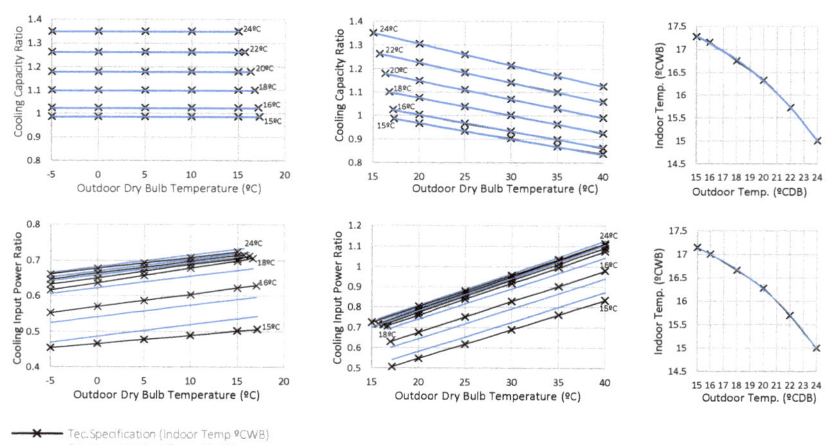

Figure A5. *Cont.*

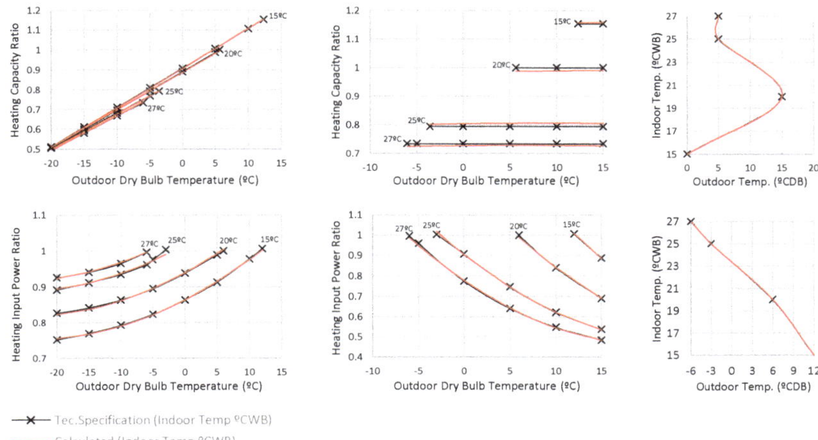

Figure A5. VRV system performance curve comparison, Technical specification displayed in black and calculated "z" value in red for heating, and blue for cooling.

For cooling, the MixedUse HP unit calculated capacity curves have a CV(RMSE) of 0.93% with an R2 of 99.4%, while its electric input has a CV(RMSE) 7.07% with an R2 of 94.5%. For heating, the calculated curve has a CV(RMSE) of 0.15% with an R2 above 99%. Its COP curve has a CV(RMSE) of 0.19% with an R2 above 99%; see Figure A6.

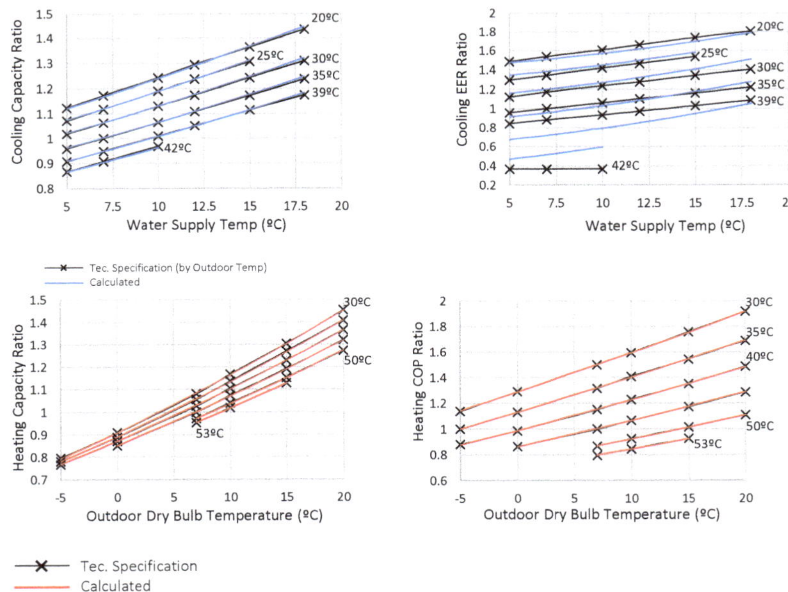

Figure A6. Heat pump performance curve comparison, technical specification displayed in black and calculated "z" value in red for heating, and blue for cooling.

Appendix C.2. MixedUse HVAC Calibration Results

Table A7. Results of VRV system calibration for cooling and heating periods, 2019–2020.

Index	ASHRAE Thresholds	Cooling Energy	Average Indoor	Heating Energy	Average Indoor
(Hourly)		Consumption	Temperature (Main Building)	Consumption	Temperature (Main Building)
DATE	—	May–October 2019	May–October 2019	November 2019– April 2020	November 2019– April 2020
NMBE (%)	within ±10%	9.49%	0.16%	17.29%	1.12%
CV(RMSE) (%)	≤ 30%	27.57%	1.04%	27.57%	2.75%
R2 (%)	≥ 75%	76.17%	95.50%	66.50%	97.39%

Table A8. Results of HP system calibration for cooling and heating periods, 2019–2020.

Index	ASHRAE Thresholds	Cooling Energy	TZ-05 Indoor	HEATING Energy	TZ-05 Indoor
(Hourly)		Consumption	Temperature (Atrium)	Consumption	Temperature (Atrium)
DATE	—	July–October 2019	July–October 2019	November 2019– April 2020	November 2019– April 2020
NMBE (%)	within ±10%	7.37%	−0.02%	−7.42%	5.31%
CV(RMSE) (%)	≤30%	27.41%	1.27%	27.21%	8.22%
R2 (%)	≥75%	78.89%	94.51%	76.49%	85.26%

References

1. Global Alliance for Buildings and Construction. *2019 Global Status Report for Buildings and Construction: Towards a Zero-Emission, Efficient and Resilient Buildings and Construction Sector*; Technical Report, International Energy Agency and the United Nations Environment Programme; Global Alliance for Buildings and Construction: Berlin, Germany, 2019.
2. IEA. *Tracking Buildings 2020*; Technical Report; IEA: Paris, France, 2020. Available online: https://www.iea.org/reports/tracking-buildings-2020 (accessed on 14 April 2021).
3. O'Neill, Z.; Li, Y.; Williams, K. HVAC control loop performance assessment: A critical review (1587-RP). *Sci. Technol. Bulit Environ.* **2017**, *23*, 619–636. [CrossRef]
4. Behrooz, F.; Mariun, N.; Marhaban, M.; Mohd Radzi, M.; Ramli, A. Review of Control Techniques for HVAC Systems—Nonlinearity Approaches Based on Fuzzy Cognitive Maps. *Energies* **2018**, *11*, 495. [CrossRef]
5. Mařík, K.; Rojíček, J.; Stluka, P.; Vass, J. Advanced HVAC Control: Theory vs. Reality. *IFAC Proc. Vol.* **2011**, *44*, 3108–3113. [CrossRef]
6. Oldewurtel, F.; Parisio, A.; Jones, C.N.; Morari, M.; Gyalistras, D.; Gwerder, M.; Stauch, V.; Lehmann, B.; Wirth, K. Energy efficient building climate control using Stochastic Model Predictive Control and weather predictions. In Proceedings of the 2010 American Control Conference, Baltimore, MD, USA, 30 June–2 July 2010; IEEE: Piscataway, NJ, USA, 2010; pp. 5100–5105. [CrossRef]
7. Ma, Y.; Borrelli, F.; Hencey, B.; Coffey, B.; Bengea, S.; Haves, P. Model Predictive Control for the Operation of Building Cooling Systems. *IEEE Trans. Control. Syst.* **2012**, *20*, 796–803. [CrossRef]
8. Moroşan, P.D.; Bourdais, R.; Dumur, D.; Buisson, J. Building temperature regulation using a distributed model predictive control. *Energy Build.* **2010**, *42*, 1445–1452. [CrossRef]
9. Schubnel, B.; Carrillo, R.; Taddeo, P.; Canal Casals, L.; Salom, J.; Stauffer, Y.; Alet, P.J. State-space models for building control: How deep should you go? *J. Build. Perform. Simul.* **2020**, *13*, 707–719. [CrossRef]
10. Tanaskovic, M.; Sturzenegger, D.; Smith, R.; Morari, M. Robust Adaptive Model Predictive Building Climate Control. *IFAC-PapersOnLine* **2017**, *50*, 1871–1876. [CrossRef]
11. Chen, B.; Cai, Z.; Bergés, M. Gnu-RL: A Precocial Reinforcement Learning Solution for Building HVAC Control Using a Differentiable MPC Policy. In Proceedings of the 6th ACM International Conference on Systems for Energy-Efficient Buildings, Cities, and Transportation, BuildSys '19, New York, NY, USA, 13–14 November 2019; Association for Computing Machinery: New York, NY, USA, 2019; pp. 316–325. [CrossRef]
12. Wei, T.; Wang, Y.; Zhu, Q. Deep Reinforcement Learning for Building HVAC Control. In Proceedings of the 54th Annual Design Automation Conference 2017, Austin, TX, USA, 18–22 June 2017. [CrossRef]

13. Schubnel, B.; Carrillo, R.E.; Alet, P.J.; Hutter, A. A Hybrid Learning Method for System Identification and Optimal Control. *IEEE Trans. Neural Netw. Learn.* **2020**, 1–15. [CrossRef] [PubMed]

14. Drgoňa, J.; Arroyo, J.; Cupeiro Figueroa, I.; Blum, D.; Arendt, K.; Kim, D.; Ollé, E.P.; Oravec, J.; Wetter, M.; Vrabie, D.L.; et al. All you need to know about model predictive control for buildings. *Annu. Rev. Control* **2020**, *50*, 190–232. [CrossRef]

15. Maddalena, E.T.; Lian, Y.; Jones, C.N. Data-driven methods for building control—A review and promising future directions. *Control Eng. Pract.* **2020**, *95*, 104211. [CrossRef]

16. Crawley, D.B.; Pedersen, C.O.; Lawrie, L.K.; Winkelmann, F.C. EnergyPlus: Energy Simulation Program. *ASHRAE J.* **2000**, *42*, 49–56.

17. Mattsson, S.E.; Elmqvist, H. Modelica—An International Effort to Design the Next Generation Modeling Language. *IFAC Proc. Vol.* **1997**, *30*, 151–155. [CrossRef]

18. Lukianykhin, O.; Bogodorova, T. ModelicaGym: Applying Reinforcement Learning to Modelica Models. In Proceedings of the 9th International Workshop on Equation-Based Object-Oriented Modeling Languages and Tools, EOOLT '19, Berlin, Germany, 5 November 2019; ACM: New York, NY, USA, 2019; pp. 27–36. [CrossRef]

19. Moriyama, T.; De Magistris, G.; Tatsubori, M.; Pham, T.H.; Munawar, A.; Tachibana, R. Reinforcement Learning Testbed for Power-Consumption Optimization. In *Methods and Applications for Modeling and Simulation of Complex Systems*; Li, L., Hasegawa, K., Tanaka, S., Eds.; Springer: Singapore, 2018; pp. 45–59.

20. Blum, D.; Jorissen, F.; Huang, S.; Chen, Y.; Arroyo, J.; Benne, K.; Li, Y.; Gavan, V.; Rivalin, L.; Helsen, L.; et al. Prototyping The BOPTEST Framework for Simulation-Based Testing of Advanced Control Strategies in Buildings. In Proceedings of the 16th International Conference of IBPSA, Rome, Italy, 2–4 September 2019; pp. 2737–2744. [CrossRef]

21. Vázquez-Canteli, J.R.; Kämpf, J.; Henze, G.; Nagy, Z. CityLearn v1.0: An OpenAI Gym Environment for Demand Response with Deep Reinforcement Learning. In Proceedings of the 6th ACM International Conference on Systems for Energy-Efficient Buildings, Cities, and Transportation, BuildSys '19, New York, NY, USA, 13–14 November 2019; Association for Computing Machinery: New York, NY, USA, 2019; pp. 356–357. [CrossRef]

22. Brockman, G.; Cheung, V.; Pettersson, L.; Schneider, J.; Schulman, J.; Tang, J.; Zaremba, W. *OpenAI Gym*; Technical Report; OpenAI: San Francisco, CA, USA, 2016.

23. Blochwitz, T.; Otter, M.; Åkesson, J.; Arnold, M.; Clauss, C.; Elmqvist, H.; Friedrich, M.; Junghanns, A.; Mauss, J.; Neumerkel, D.; et al. Functional Mockup Interface 2.0: The Standard for Tool independent Exchange of Simulation Models. In Proceedings of the 9th International Modelica Conference, Munich, Germany, 3–5 September 2012; pp. 173–184. [CrossRef]

24. Ruiz, G.R.; Bandera, C.F.; Temes, T.G.A.; Gutierrez, A.S.O. Genetic algorithm for building envelope calibration. *Appl. Energy* **2016**, *168*, 691–705. [CrossRef]

25. Deb, K.; Pratap, A.; Agarwal, S.; Meyarivan, T. A fast and elitist multiobjective genetic algorithm: NSGA-II. *IEEE Trans. Evol. Comput.* **2002**, *6*, 182–197. [CrossRef]

26. Ruiz, G.R.; Bandera, C.F. Validation of calibrated energy models: Common errors. *Energies* **2017**, *10*, 1587. [CrossRef]

27. Fuentes, E.; Salom, J. Validation of black-box performance models for a water-to-water heat pump operating under steady state and dynamic loads. *E3S Web. Conf.* **2019**, *111*, 01068. [CrossRef]

28. Tang, C.C. Modeling Packaged Heat Pumps in a Quasi-Steady State Energy Simulation Program. Master's Thesis, Oklahoma State University, Stillwater, OK, USA, 2005. Available online: https://hvac.okstate.edu/sites/default/files/pubs/theses/MS/27-Tang_Thesis_05.pdf (accessed on 14 April 2021).

29. Wetter, M. Modelica Buildings library. *J. Build. Perform. Simul.* **2014**, *7*, 253–270. [CrossRef]

30. Swiss Society of Engineers and Architects. Klimadaten für Bauphysik, Energie- und Gebäudetechnik, SIA 2028:2010. Available online: http://shop.sia.ch/normenwerk/architekt/sia%202028/f/2010/D/Product (accessed on 14 April 2021).

Review

How Can Existing Buildings with Historic Values Contribute to Achieving Emission Reduction Ambitions?

Selamawit Mamo Fufa [1,*], Cecilie Flyen [1] and Anne-Cathrine Flyen [2]

[1]   Department of Architectural Engineering, SINTEF Community, P.O. Box 124 Blindern,
      NO-0314 Oslo, Norway; Cecilie.flyen@sintef.no
[2]   Department of Buildings, Norwegian Institute for Cultural Heritage Research, Storgata 2, 0155 Oslo, Norway;
      anne.flyen@niku.no
*   Correspondence: selamawit.fufa@sintef.no

Abstract: In line with the Paris Agreement, Norway aims for an up to 55% reduction in greenhouse gas (GHG) emissions by 2030 compared to 1990 levels and to be a low-emission society by 2050. Given that 85–90% of today's buildings are expected to still be in use in 2050, refurbishment and adaptive reuse of existing buildings can help in achieving the environmental goals. The aim of this work is to provide a holistic picture of refurbishment and adaptive reuse of existing buildings, including buildings with heritage values, seen from a life cycle perspective. The methods applied are a literature review of LCA studies and experiences from quantitative case study analysis of selected Norwegian case studies. The findings show that extending the service life of existing buildings by refurbishment and adaptive reuse has significant possibilities in reducing GHG emissions, keeping cultural heritage values, and saving scarce raw material resources. The findings show limited LCA studies, uncertainties in existing LCA studies due to variations in case-specific refurbishment or intervention measures, and a lack of transparent and harmonized background data and methodological choices. In conclusion, performing a holistic study covering the whole LCA and including socio-cultural values and economic aspects will enable supporting an argument to assert the sustainability of existing buildings.

Keywords: embodied emission; existing building; LCA; heritage value; socio-cultural value

Citation: Fufa, S.M.; Flyen, C.; Flyen, A.-C. How Can Existing Buildings with Historic Values Contribute to Achieving Emission Reduction Ambitions? *Appl. Sci.* **2021**, *11*, 5978. https://doi.org/10.3390/app11135978

Academic Editors: Elmira Jamei and Zora Vrcelj

Received: 25 May 2021
Accepted: 21 June 2021
Published: 27 June 2021

Publisher's Note: MDPI stays neutral with regard to jurisdictional claims in published maps and institutional affiliations.

1. Introduction

Adopted by 195 countries in December 2015, the Paris Agreement was a major step forward for a global action plan in mitigating climate change [1]. The agreement aims to keep global warming below 2 °C above pre-industrial levels, and to continue efforts to limit global warming to below 1.5 °C to prevent the adverse effects of climate change. The EU has been at the forefront of international efforts to fight climate change by setting ambitious energy and climate targets towards 2020 and 2030, and by progressively working towards a climate-neutral EU by 2050 [2]. The climate ambitions for 2030 have been revised in line with the Paris Agreement goals, with an ambition of cutting GHG emissions by at least 55% compared to 1990 levels [3]. The previous EU target was a 40% GHG emission cut while improving energy efficiency by at least 32.5% and the share of renewable energy by at least 32%. The EU Green Deal, proposed by the EU Commission in December 2019, sets a road map on how to achieve the newly set climate goals [4].

Despite the progress in climate policy and ambitions in many countries, global GHG emissions continue to grow to the highest ever, as recorded in 2019 [5]. Buildings are responsible for about 40% of the EU's total energy consumption, and for 36% of its greenhouse gas emissions from energy [6]. More than 220 million building units (85% of the EU's building stock) were built before 2001, and 85–95% of existing buildings today will still be standing in 2050 [6]. However, almost 75% of those existing buildings are energy-inefficient, and the annual energy renovation rate is lower than 1% [6]. The Norwegian building stock is expected to follow the same trend. The construction industry is seen as essential in the transition to a circular economy, as it consumes around 40% of the total material resources and generates over one third of the total waste in the world [7]. The Norwegian national statistical data show that waste from construction, refurbishment, and demolition (1.95 million tons in 2019) represented around 26% of the total national waste, an increase by 5.6% from 2018 [8]. Of this, about 76.3% of the waste is from demolition and refurbishment activities, and only 46% of the total waste was recycled in 2018, which is a decrease of 8% from the previous year.

To achieve the 55% emission reduction target by 2030, the EU should reduce buildings' greenhouse gas emissions by 60%, their final energy consumption by 14% and energy consumption for heating and cooling by 18%. The COVID-19 crisis has clearly demonstrated the importance buildings play in our daily life, and also the unique opportunities that lie in refurbishment in terms of rethinking, redesigning, and modernizing existing buildings into more energy-efficient and less material-intensive buildings and sustaining economic recovery [6]. Further, this is sustainable from a holistic and cultural heritage perspective, due to upholding and continuing the socio-cultural values the existing buildings, in general, and heritage buildings, specifically, represent.

With the growing carbon spike and climate change impacts, refurbishment and adaptive reuse of existing buildings, and heritage buildings, can be conducive to taking immediate action and achieving various emission reduction goals. The EU Green Deal started a new renovation initiative in 2020, the "renovation wave", with a target to doubling the current public and private renovation rate in the next 10 years [9]. Applying circularity principles would help conserve scarce raw material resources, reduce GHG emissions associated with carbon-intensive production processes (i.e., cement, steel, and glass), and boost economic crises resulting from the pandemic. The evaluation of the environmental performance of existing buildings in a life cycle perspective, using transparent and harmonized methods, will enable avoiding problem shifting, and evaluating and following up the fulfilment of emission reduction ambitions.

This paper presents a summary of the main lessons learnt from the life cycle assessment of four Norwegian case studies to shed light on the gaps in the above-mentioned challenges and opportunities. This paper begins by outlining the state-of-the-art literature, followed by a short description of four case studies and the methodology used for evaluating and discussing these. The LCA results of the case studies are presented, and the findings are discussed, in order to provide background for lessons learnt from each case study. After highlighting the needs of future research activities, final remarks are drawn in the conclusion.

2. Literature Review

This section provides relevant studies focusing on definitions of terminologies (to avoid misinterpretation), and refurbishment and adaptive reuse of existing buildings and life cycle assessment are presented.

2.1. Definitions

The terms renovation, refurbishment, restoration, retrofitting, upgrading, and adaptive reuse are often used interchangeably and inconsistently. Shahi et al. [10] conducted a literature review of terminologies related to building adaptation and developed a framework by summarizing the definition of terminologies into two main categories:

- Refurbishment: The process of improvement or modification of an existing building, through maintenance, repair, improvement/upgrading of existing systems, and incorporation of energy efficiency measures, in order to make it fulfil current building standards for existing use (without change of use). Refurbishment is categorized into the following:

- o Retrofitting: addition or upgrading of building envelope, HVAC systems, energy efficiency, and renewable energy sources to improve the energy use and efficiency of an existing building.
  - o Renovation: replacing or fixing old components to increase energy efficiency or remodeling of the interior layout of an existing building to improve the esthetic appearance or interior design.
  - o Rehabilitation: repairing, changing, or adding to damaged structures, deteriorating envelopes, openings, and HVAC systems.

- Adaptive reuse: Extending the service life of historic, old, obsolete, and abandoned buildings through reusing an existing building for a different use or reusing the building materials and structures for a different use. The two aspects of adaptive reuse of existing buildings are described as follows:

  - o Conversion: adaptive reuse of obsolete and abandoned buildings, which are not used anymore or do not satisfy their users, through retrofitting, renovation, and rehabilitation of the building envelope and structures by changing the original function of the building partially or entirely.
  - o Material reuse: recover and reuse of materials from an existing building through partial repair or refurbishment.

In this paper, the terms refurbishment, when discussing rehabilitation, retrofitting, renovation, and upgrading, and, accordingly, adaptive reuse, when discussing conversion and material reuse, are considered. Historic buildings, or heritage buildings, are referred to as buildings with heritage significance (not necessarily considered as listed or protected) due to their historical, architectural, symbolic, social, or cultural values [11].

In this paper, a historic building is referred to as an existing building, not necessarily listed or protected, but worth saving. Within the preservation of cultural heritage, renovation measures generally entail interventions that undermine heritage values. Thus, measures incorporating changes in construction details, materials, and appearances of the façades are most often not allowed, or wanted, by heritage authorities [12].

2.2. Refurbishment and Adaptive Reuse

Traditionally, considerate refurbishment of existing buildings, and historic and heritage buildings in particular, has not been a priority. Decisions to keep or demolish existing buildings often lead to several uncertainties as summarized by the Norwegian Green Building Council [13]: (1) refurbishment is more costly than demolition and constructing new buildings; (2) only new buildings can be environmentally certified; (3) existing buildings are not area-efficient; (4) refurbished buildings have difficulties in satisfying modern indoor climate requirements; and (5) refurbished buildings often have poor daylight qualities. The level of the refurbishment is also dependent on the building's historic and heritage values.

Refurbished buildings do, however, provide huge immediate environmental benefits [14,15]. Refurbishment of cultural heritage buildings, with the use of existing resources where possible, rather than dependence on new resources, is the most sustainable course of action. Foster [16] pointed out how multiple analyses in recent research concur that adaptive reuse of existing buildings is beneficial for the environment, and, further, that current research upholds the environmental benefits from adaptive reuse of buildings. In addition, refurbishing and adaptively reusing under-utilized or neglected buildings can revive neighborhoods whilst achieving environmental benefits, and such buildings embody the local socio-cultural and historic characteristics that define communities [16]. Foster [16] does, however, maintain that such benefits are not widely adopted in practice.

Energy efficiency measures in the transformation and reuse of heritage buildings might also lead to increased energy needs, due to a change in use and user needs [17]. Fouseki and Cassar [18] proved that user behavior is often more important than the type of technological selection in energy efficiency measures, both considering the amount of energy spent and the way the building is used. Most of such studies have mainly targeted modern buildings, materials, and constructions. Policy measures for cultural heritage and sustainability should be communicated to achieve public understanding concerning user behavior, energy efficiency measures, and heritage values [19]. Berg et al. [20] highlighted the significant impact and potential of non-intrusive energy efficiency measures and integrated bottom-up refurbishment processes providing for better management.

The building stock represents an important cultural and material resource, where some buildings are of heritage significance due to their, e.g., historic, architectural, social, and cultural values [11]. Many buildings subjected to upgrading or restoration may have been built for other purposes, and for other comfort requirements, which is presently the case. Often, such requirements will conflict with preserving inherent heritage values [19].

The Paris Agreement and the United Nations' Sustainable Development Goals (SDG) provide explicit recognition of the role of cultural heritage in promoting climate change mitigation and adaptation measures [21]. Meeting the need for environmental impact reduction and cost-effective energy-efficient solutions while, at the same time, safeguarding a building's cultural heritage value does, however, present a clear technical challenge [22]. Challenges of balancing different needs, limitation of tools for investigation of energy efficiency measures, and a lack of knowledge of historic and heritage buildings are considered as key challenges when implementing energy efficiency strategies in such buildings [23].

Nastasi and Matteo [24] argued that focusing on incorporating a renewable energy supply system instead of modification of the building envelope or installation of a new HVAC system is a possible solution for preserving existing buildings with historic significance. Studies have shown the possibilities of functional improvement in historic buildings due to technological evolutions, such as a district heating system dominated by biomass [25], integration of hydrogen into natural gas as a partial fuel substitution [24], and HVAC systems [26].

Development of best practice methods, tools, and guidelines with quantitative analyses which support qualitative study of existing buildings with historic values at a large scale (districts and cities rather than the individual building level) will enable filling in the current knowledge gap in this area [22]. Ruggeri et al. [27] developed a score-driven decision making model which can be used to plan and manage energy retrofitting measures applied to cultural heritage and historic buildings. The model is described as a combination of quantitative and qualitative parameters, including energy savings and costs, building conservation aspects measured as a restoration score, and their impact on indoor comfort. Bertolin and Loli [28] developed a zero emission refurbishment (ZER) tool to support decision-makers on maintenance and adaptive intervention principles, and to analyze potential emissions and energy consumptions (that should be compensated with onsite renewables) of interventions of historic buildings.

Since the proportion of existing buildings is large in relation to the number of new, energy-efficient buildings, and, furthermore, since many existing buildings have lower energy efficiency, there is a need to assess the effect of energy-efficient measures in the existing building stock. Such assessments must be viewed in the context of the value added, other upgrading and maintenance needs, changing comfort requirements, and the effect of the measures in relation to cost and emission savings. This demonstrates that there is a need to develop strategies addressing both emission and adaptation considerations for historic and heritage buildings, regardless of ownership. EN 16883 [11] states the importance of evaluating the environmental performance of heritage buildings from a life cycle perspective when making a decision on energy performance improvement measures.

2.3. Life Cycle Thinking

Life cycle assessment has been used as a decision support tool to evaluate the environmental performance of buildings. It enables identifying hot spots, avoiding problem shifting, and making informed decisions. There have been many advancements in developing and harmonizing building LCA standards (e.g., ISO 21931 [29] for LCA principles and framework for assessment of environmental performance of construction works, EN

15897 [30] for the assessment of environmental performance of buildings, and the national Norwegian standard NS 3720 [31]), and in LCA results communication methods (e.g., EPDs, and green building rating systems such as BREEAM) and international research activities (such as the International Energy Agency's Energy in Building and Communities program (IEA EBC) Annex 57 [32], Annex 72 [33]) on assessing and communicating life cycle-related impacts of buildings.

Life cycle-based national GHG emission benchmarks or reference values are receiving more attention in different countries, and there are on-going discussions on the possibility of legal bindings [34]. The lack of harmonized background data and methodology choices used in different studies is considered as a major challenge in the utilization of LCA studies to establish benchmarks [34–36]. Most existing benchmarks also provide aggregated values for both new and existing buildings, mainly due to a lack of LCA studies from existing buildings and heritage buildings.

As with other Nordic countries, in Norway, there is no specific legal requirements for building LCA in the national legislation yet [37]. However, public building owners (e.g., the Norwegian government agency for administration of public buildings in Norway, Stastbygg), large municipalities (e.g., Oslo), initiatives, and projects (e.g., Futurebuilt, FME ZEB, FME ZEN) have been actively promoting the use of building LCA in Norwegian buildings for the past 15 years [37,38].

Stricter legislation towards the implementation of energy efficiency measures and decarbonization of national energy mixes (through use of renewable energy sources) leads to a reduction in operational energy use [39]. However, these measures also increase the embodied emissions from: the initial embodied emissions (from manufacturing (A1–A3 life cycle stages)), transport to construction site (A4), construction installation activities (A5), recurrent embodied emissions (from use phase activities, B1–B5), and end of life embodied emission (C1–C4) (Figure 1).

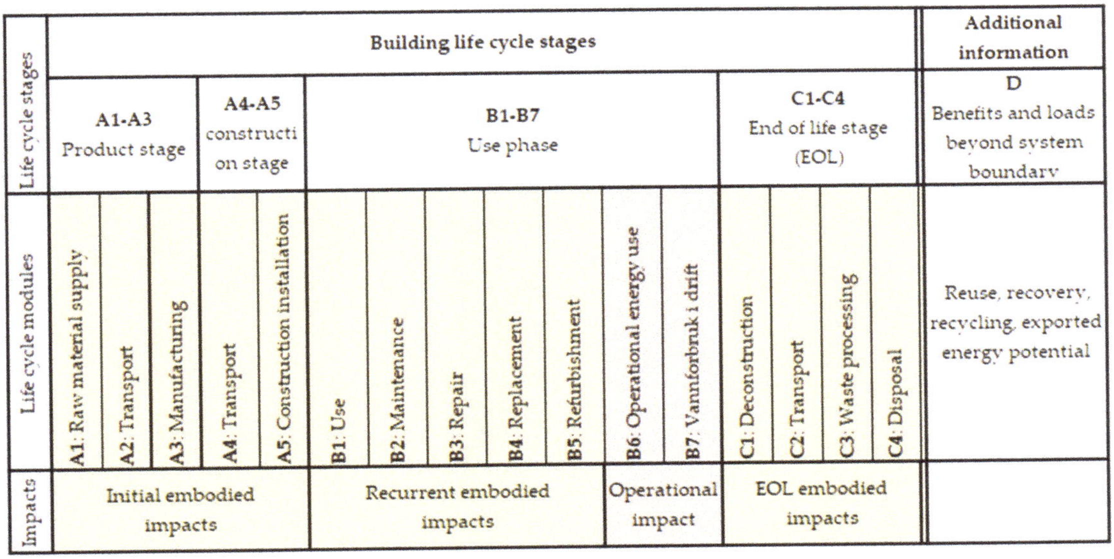

Life cycle stages	Building life cycle stages				Additional information
	A1-A3 Product stage	**A4-A5** construction stage	**B1-B7** Use phase	**C1-C4** End of life stage (EOL)	**D** Benefits and loads beyond system boundary
Life cycle modules	A1: Raw material supply / A2: Transport / A3: Manufacturing	A4: Transport / A5: Construction installation	B1: Use / B2: Maintenance / B3: Repair / B4: Replacement / B5: Refurbishment / B6: Operational energy use / B7: Vannforbruk i drift	C1: Deconstruction / C2: Transport / C3: Waste processing / C4: Disposal	Reuse, recovery, recycling, exported energy potential
Impacts	Initial embodied impacts		Recurrent embodied impacts / Operational impact	EOL embodied impacts	

Figure 1. System boundaries in building LCA in relation to life cycle stages, modules, and types of impacts (embodied and operational). Adopted from EN 15978 [30] and Moncaster et al. [40].

Most existing studies mainly cover the production (A1–A3), replacement (B4), and operational energy use phases (B6) [40–42]. The embodied impacts from the transport (A4), construction installation (A5), and end of life (C1–C4) phases are often considered as low, and those life cycle stages are either excluded or calculated based on very rough assumptions. However, on-going fossil-free, emission-free, waste-free, reuse of building products, and other circular economy activities show the importance of those phases. For example, the GHG emission from construction site activities is estimated to be 5–10% [43]. A recent report from Bellona estimated that construction-related transport accounts for 1% of Norway's total emissions [44]. The report also shows a potential GHG emission reduction by up to 50% by efficient transport of goods to and from Norwegian construction sites through better planning and higher utilization of the vehicle capacity. Adding the impacts from the construction use phase (B2–B5) and end of life stages can help to demonstrate the significance of the emission reduction potential from refurbishment of existing buildings.

The findings from the Norwegian ZEB case studies (four new buildings, two concept studies, and one refurbishment building) showed that the embodied impact of ZEBs ranges between 55 and 87% from A1 to A3 and B4, 2 and 15% from A4 to A5, and ca 8% from C1 to C4 [42], whilst the operational impact represented 14–42%. The LCA study from the IEA EBC Annex 57 (Annex 57) on 80 building case studies (where 11 refurbished buildings were included) found that the production phase dominates total GHG emissions, with 64% of GHG emissions, followed by replacements at 22% and end of life at 14% [40].

Existing studies also indicate the potential environmental benefits from refurbishment of existing buildings in the range between 4 and 74% depending on the scenarios considered in the studies. Assefa and Amber [45] indicated a 28–33% impact reduction from the selective deconstruction and reuse of a thirteen-story library tower at the University of Calgary, in Western Canada, in seven environmental impact categories assessed (eutrophication potential, smog potential, global warming potential, fossil fuel consumption, human health criteria, acidification potential, and ozone depletion). Assefa and Amber also pointed out the importance of comparative assessments for reuse vs. new construction regardless of the challenges due to the uniqueness of different buildings and their locations.

Hasik et al. [46] showed 36–75% impact reductions across six environmental impact indicators (acidification potential, eutrophication potential, global warming potential, ozone depletion potential, smog formation potential, and non-renewable energy demand) when refurbishment of an office building was compared to new construction. Hasik et al. [46] also addressed the challenges related to the lack of a clear system boundary description regarding life cycle modules included in existing buildings and refurbishment projects. The study performed by Preservation Green Lab [47] on six building typologies from four US cities showed about a 4 to 46% environmental impact saving from renovation and reuse of existing building rather than demolishing and constructing new buildings. Preservation Green Lab [47] discussed the potential dependency of impacts from the energy performance of upgrading associated with material choices. Eskilsson [48] discussed the dependency of operational energy use on the sources of energy and the associated emission factors and highlighted the importance of embodied GHG emissions.

Even if refurbishment and adaptive reuse of existing buildings have been suggested for reducing embodied impacts [46], only a few LCA studies focus on the evaluation of the environmental performance of refurbishment and adaptive reuse of the existing building stock [49], and even fewer studies consider embodied impacts and evaluate the contribution of cultural and heritage values [16,22,46,50]. For example, from the 80 studies evaluated under Annex 57, only 11 were existing buildings [40]. The results from the first LCA benchmark study conducted in Norway also show similar findings, as from over 120 studies collected, only 13 were for existing buildings, and only 2 of the 13 studies presented the results from a historic building. The results in both Annex 57 and the Norwegian studies indicate the impact from refurbishment projects, which is under half that of new building projects. Developing LCA benchmarking for existing buildings, based on a harmonized methodology, will increase the transparency and repeatability of LCA results and enable different user groups to make informed decisions.

Interventions to reduce emissions will often affect heritage values if special considerations are not taken into account [20]. When restoring or preserving legally protected buildings in Norway, value assessments always form the basis for measures permitted by the cultural heritage authorities [12]. In Norway, the cultural heritage authorities also want buildings without formal protection to be treated in accordance with their heritage values [51].

3. Case Studies

The methodology in this study includes a quantitative review used to critically evaluate and analyze different Norwegian case studies to provide a quantitative answer to research gaps identified under Chapter 2. Four Norwegian case studies, all refurbishment projects where LCA studies were conducted, were selected for the analysis. The calculation was performed using a functional unit of 1 m^2 and a building lifetime of 60 years. LCA practitioners have used either klimagassregnskap.no (KGR.no, the previous Norwegian GHG emission calculation tool, using the Ecoinvent database and EPDs as background data), SimaPro (commercial LCA software using the ecoinvent database), OneClick LCA (commercial LCA tool which replaced KGR.no in 2018), or the ZEB tool (Excel-based GHG emission calculation tool developed by the Norwegian ZEB Centre, using the Ecoinvent database and EPDs as background data) for GHG emission calculations. The calculations follow the LCA methodology outlined by LCA standards, NS 3720, EN 15978, and/or ISO 14040/44.

The Norwegian case studies consist of one residential building: Villa Dammen, and three office buildings: Powerhouse Kjørbo, Bergen city hall, and Statens bygg Vadsø, bygg B. Villa Dammen, Bergen city hall, and Statens bygg Vadsø, bygg B (herein referred to as Statens bygg Vadsø) are buildings with heritage significance. The heritage values were not considered directly in the LCA model when assessing the life cycle analysis as this type of value is not possible to consider in the existing models. However, heritage values were considered when choosing measures to reduce emission values. Powerhouse Kjørbo, an office building from the 1980s, was selected as a case study as it was the first refurbishment project in the world which produces more energy than it consumes. The summary of the general information about the case studies and the results from the life cycle assessment is shown in Table 1.

Table 1. Summary of case studies considered in this study.

	General Information			
Name of the building	Villa Dammen	Statens hus Vadsø	Bergen city hall	Powerhouse Kjørbo
References	[52,53]	[54]	[55]	[56]
Typology	Residential	Office	Office	Office
Location	Moss	Vadsø	Bergen	Sandvika
Heated floor area, BRA (m^2)	117	3460 (for new building scenario) [1] 4297 (for refurbishment scenario) [1]	10,756	5180
Service life (year)	60	60	60	60
No. of floors (no.)	2	4	14	4–5
Construction period (year)	1936	1963	1974	1980
Renovation period	2014–2015	2021–2024	2019	March 2013–February 2014
Emission factor for electricity and for other energy sources ($kgCO_2eq/kWh$)	0.132 (ZEB factor), 0.321 (for paraffin), 0.014 (for biofuel)	0.13 (EU factor)	0.195 (EU factor) 0.11 (for district heating)	0.132 (ZEB factor)
Building phase	As-built phase	Design phase	Design phase	As-built phase
Refurbishment or adaptive reuse	Refurbishment	Refurbishment	Refurbishment	Refurbishment
Parts of the building rehabilitated	Sealing around windows and doors, floor and roof insulation, heating system	Interior walls, floor, ceiling, insulation of roof (300 mm) and outer wall (150 mm), and scenarios for technical installation	Interior walls, floor, ceiling, façade, and concrete elements	Outer laminated glass façade was recycled and used internally, and the exterior walls were replaced, re-insulation of roof and exterior walls of the basement
Life cycle modules covered (reported)	A1–A3, B4, C1–C4 (material use), and B6 (operational energy use)	A1–A5, B4, C1–C4 (material use), A5 (installation), and B6 (operational energy use)	A1–A5, B4–B5, C1–C4 (material use), B6 (operational energy use), B8 [2], D [2]	A1–A5, B4, C1–C4 (material use), B6 (operational energy use)
LCA tool	SimaPro	OneClick LCA	OneClick LCA	ZEB tool
Emission factors for electricity considered for scenarios analysis ($kgCO_2eq/kWh$)	0.132 (ZEB factor [3])	0.13 (EU28 + NO factor) 0.0128 (NO factor)—both factors calculated following NS3720, production from 2015–2017	0.195 (EU28 + NO factor) 0.024 (NO factor)—both factors calculated following NS3720	0.132 (ZEB factor [3])
Scenarios used in the analysis and (codes)	(1) Without refurbishment (without refurb.), after refurbishment (Refurb.), demolition and building new (New) (2) Measured and calculated operational energy use	(1) New building as reference (New) and after refurbishment (Refurb.), (2) EU factor vs. NO factor for electricity, (3) With and without technical installation (tech. install.)	(1) New building as reference (New) and after refurbishment (Refurb.), (2) EU factor vs. NO factor for electricity	(1) Measured and calculated operational energy use

Table 1. *Cont.*

| | GHG Emission Results of Case Studies Per Life Cycle Module (kgCO$_2$eq/m^{2}BRA/yr.) | | | | | | | |
| | Villa Dammen | | Statens Hus Vadsø | | Bergen City Hall | | Powerhouse Kjørbo | |
	Refurb.	New	Refurb.[4]	New [4]	Refurb.	New	Calculated	Measured
A1–A3	0.40	4.60	2.74 [5]	6.24 [5]	1.10		3.77	3.77
A4–A5			0.11 [5]	0.28 [5]	0.20	5 [6]	0.25	0.25
B4	0.90	1.70			0.40 [6]		1.82	1.82
B6	18.00	11.60	13.21 [5]	7.85 [5]	19.00	18.30	6.54	5.82
C1–C4	0.90	0.70			0.03		0.74	0.74
D	NA [7]	NA [7]	NA [7]	NA [7]	NA [7]	NA [7]	−5.82	−5.70
Total GHG emission ((kgCO$_2$eq/m^{2}BRA/yr)	20.2	18.6	16.05	14.37	20.73	23.3	7.30	6.70

[1] The GHG emission results given per gross floor area of new (3680 m^2) and refurbished (4555 m^2) buildings are recalculated to per heated floor area (BRA). [2] The results from B8 (transport in use phase, according to NS 3720) and D are not included in this paper. [3] This is the emission factor used in the Norwegian ZEB pilots [57]. The emission factor is yearly averaged based on an assumption of a future scenario of a fully decarbonized European grid by the end of 2050. [4] The results include A1–A5, B4–B5, and C1–C4 values. [5] The results include A1–A5, B4–B5, and C1–C4 values. [6] The result of B4 also includes B5 values. [7] Not Available (NA).

4. Lessons Learned from Norwegian Case Studies

This section presents and discusses the findings from the LCA results of each case study focusing on the share between operational and embodied emissions and factors affecting the LCA results. Moreover, the limitations and topics for further research are discussed.

4.1. Embodied and Operational GHG Emissions

Figure 2 shows the embodied and operational GHG emission results from the four case studies. The results show that the operational GHG emissions of the refurbishment range from 50 to 91%, whilst the operational GHG emissions from new buildings range between 55 and 79%. For the refurbishment, the lowest value (50%) is from Powerhouse Kjørbo, where the ZEB factor is considered. The highest value (91%) is from Bergen city hall, where the EU factor is used. For the new building, the lowest operational GHG emission (55%) is from Statens hus Vadsø, where the EU factor and the impact from the technical installations are considered. Meanwhile, the highest operational GHG emission (79%) is from Bergen city hall.

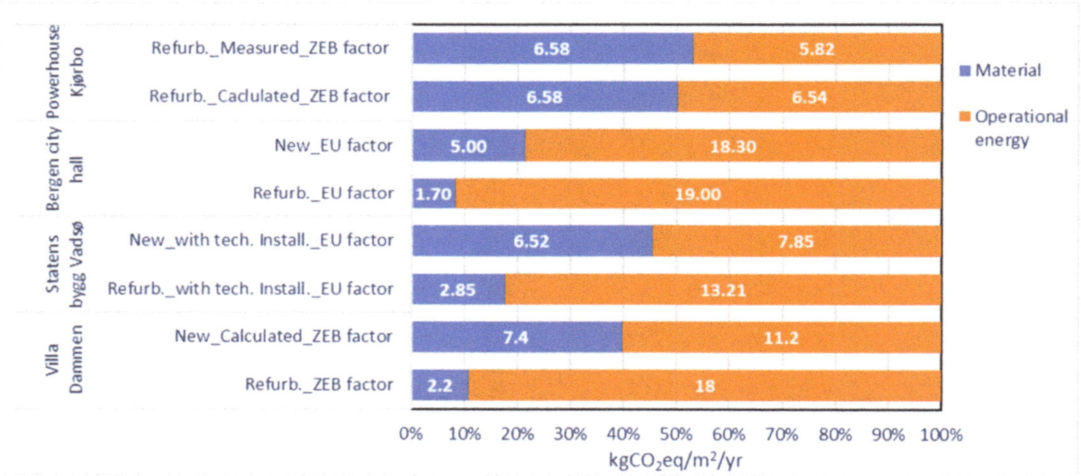

Figure 2. Embodied and operational GHG emission results.

The embodied GHG emissions of the refurbishment range from 8 to 50%, whilst the embodied GHG emissions from new buildings range between 21 and 45%. For the refurbishment, the lowest value (8%) is from Bergen city hall, where the operational GHG emissions dominated due to the use of the EU factor. The highest value for the refurbishment (50%) is from Powerhouse Kjørbo, where the energy efficiency measures were aiming at refurbishing the building to ZEB-COME. For the new building, the lowest embodied GHG emissions (21%) are from Bergen city hall, where the EU factor is considered. Meanwhile, the highest embodied GHG emissions (45%) are from Statens hus Vadsø, where the EU emission factor and impact from technical installations are considered.

The overall results show that the lowest embodied GHG emissions originate from the refurbishment of existing buildings compared to new buildings. A Historic London study [50] showed that the embodied GHG emissions from a new building accounted for 30% of emissions in comparison to 2% from the refurbishment of a Victorian terraced house from 1891 (which is representative of a large number of British homes). Thus, by conserving existing buildings, the embodied GHG emissions that are inherent in the generation of waste during the demolition of old buildings can be reduced. Furthermore, the materials used and waste generated in the construction of new buildings and other emissions associated with transport and energy consumption linked to the construction of new buildings can be reduced.

4.2. Influence of Emission Factors

The refurbishment scenarios have lower embodied GHG emissions in Bergen city hall (8%), Villa Dammen (11%), and Statens hus Vadsø (18%), but the higher operational energy consumption leads to higher total GHG emissions from refurbishment scenarios in comparison with the new buildings. Different scenarios were considered to evaluate the influence of the emission factor, technical installations, and without refurbishment results (Figure 3).

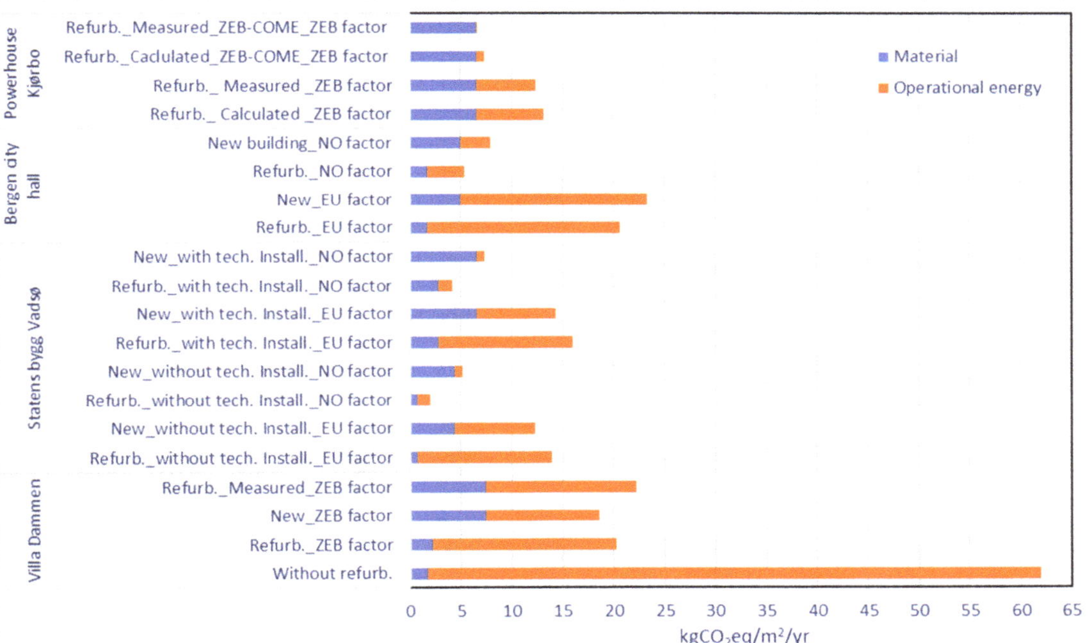

Figure 3. Scenario analysis.

In Villa Dammen and Powerhouse Kjørbo, the influence of the calculated and measured energy consumption was considered, whilst in Statens hus Vadsø and Bergen city hall, the NO and EU emission factors were considered (Figure 3). The change in these scenarios did, however, not have any impact on the embodied GHG emissions.

The operational GHG emissions from the measured energy consumption values were lower than the calculated values both in the Villa Dammen (14.8 kgCO$_2$eq/m^2/yr) and Powerhouse Kjørbo (5.82 kgCO$_2$eq/m^2/yr) refurbishment scenarios. The user's energy use behavior was considered as one of the reasons behind the lower energy consumption in the actual measured values. NS-EN 16883 [11] emphasized the importance of raising users' awareness on the influence of their behavior on energy consumption and associated costs.

Powerhouse Kjørbo is a pilot project of the Norwegian ZEB Research Centre, setting an ambition of achieving the ZEB-COME ambition level, following the step-wise Norwegian ZEB ambition level definition [57]. ZEB-COME means all emissions related to construction (C), operation (O), production and replacement of materials (M), deconstruction, transport, and disposal at end of life (E) shall be compensated for with onsite renewable energy generation. The result for the ZEB-COME scenario shows that the onsite energy generation (-5.82 kgCO$_2$eq/m^2/yr for calculated and -5.7 kgCO$_2$eq/m^2/yr for measured) compensated for 44 (for calculated) to 46% (for measured) of the total GHG emissions.

The comparative assessment of the refurbishment vs. new building scenarios in Statens hus Vadsø and Bergen town hall shows the influence of the scenarios for the choice of CO$_2$ factors on the operational and embodied GHG emissions. When using the EU emission factor, the newly built project has the lowest emissions, and when using the NO emission factor, the refurbishment project has the lowest emissions. The potential environmental benefit of refurbishment is more than half compared to a reference new building scenario. This is also in line with the findings from the IEA EBC study on 11 renovated buildings [40].

The results demonstrate that the lowest GHG emissions from operational energy are found for scenarios when the NO emission factor (lower emission factor, because of electricity generated by renewable resources) is considered. The results from Statens hus Vadsø show that when including the impacts from technical installations, the embodied impact of the new building is further increased by 67% in comparison with the refurbishment (Figure 3). Most studies excluded technical installations; however, Moncaster et al. (2019) estimated the impact of technical installations as being between 9 and 45% of the total life cycle emissions.

In addition, using the EU emission factor resulted in increasing the significance of the emissions from operational energy in comparison to embodied GHG emissions. Thus, considerations of measures that reduce operational energy demands increase the energy efficiency, and the use of renewable energy sources can lead to increases in embodied impacts. Georges et al. [58] demonstrated similar findings in the Norwegian ZEBs where the embodied GHG emissions dominate the operational energy emissions when a low emission factor is considered. The choice of emission factor for electricity affects the contribution of the embodied GHG emissions. Andersen et al. [59] argued that lower emission factors can lead to yielding more weight on embodied emission reduction measures and lower weight on operational energy reductions, and vice versa for higher emission factors. It is important to note that even if both case studies (Statens Hus Vadsø and Bergen city hall) claim to have used the EU28 + NO factor, and that the NO factor was calculated based on NS3720, the emission factors used in the two projects were different. Providing additional information about the calculation method would have increased the transparency of the scenarios. However, such detailed information was not available.

The lessons learnt from the Norwegian ZEB pilots identified the importance of implementation of energy efficiency measures, use of renewable resources, considerations of efficient materials, area reduction measures, prioritizing the use of reused and recycled materials, and the use of long-lasting, low-embodied carbon, and locally available materials to reduce potential embodied GHG emissions [42,60]. Similar findings were obtained through the analysis of IEA EBC Annex 57 case study collections [40]. This is also in line with the three core elements of circular economy and the 10R circular economy framework, which make up the core elements: (1) prioritize regenerative resources (refuse, reduce, rethink), (2) service life extension (reuse, repair, refurbish, remanufacture), and (3) use waste as a resource (repurpose, recycle, recover) [61].

4.3. Time Aspects

The most common LCA method (attributional LCA) uses static calculation perspectives assuming that the emissions occurring today have the same impacts as the emissions in the future. Considerations of emission reduction measures today and in the near future are important to limit the effects of global emissions over time. Evaluation of the time it takes for new buildings to become less carbon-intensive than refurbished buildings, the payback time, is thus the way of showing the importance of refurbishment to make informed decisions. However, the length of the payback period can vary based on several factors including depth of energy efficiency refurbishment measures, emission factors for operational energy use, carbon intensity of materials, and systems used for refurbishment and demolition [50].

Even for the EU factor, which leads to higher operational energy compared to the NO factor, the GHG emissions from the Statens hus Vadsø refurbishment scenario were lower than the new building for up to 22 years (for emissions including the technical installation) and 24 years (for emissions excluding the technical installation) of the building lifetime. This means that after 22 years, the associated lower operational energy GHG emissions will make the new build scenario more beneficial than refurbishment. On the other hand, for the NO factor, the refurbishment has lower GHG emissions than the new building (for both emissions with and without technical installation scenarios) over 60 years of the building lifetime.

In Villa Dammen, the GHG emissions for the refurbishment scenario were lower than for the new building for up to 52 years. This means that in Villa Dammen, considering the operational GHG emissions can only lead to underestimating the GHG emissions from new buildings by up to 40% over 52 years and make refurbishment a less attractive option for emission savings.

From an environmental goal perspective, refurbishment of the Statens hus Vadsø and Villa Dammen buildings will be a better option to achieve the 2030 and 2050 environmental goals. The Historic London study [50] results showed a lower GHG emission from the Victorian terrace refurbishment for up to 60 years and also showed refurbishment as the best option to achieve the 2030 and 2050 lower GHG emission goals compared to demolition and new building scenarios. The results from a Norwegian study [62] noted that refurbishment strategies are in line with the 2030 goals; however, new buildings need to implement more climate mitigation strategies to close the gap. To reach the 2050 climate goals, much more is needed to be conducted for both new and existing buildings.

This demonstrates that refurbishment of existing buildings with historic/heritage values should be favored over demolition and new buildings to achieve short- and medium-term environmental goals.

4.4. Cultural Values and Storytelling Potential

The cultural values of a building often limit the scope of the energy efficiency measures. In Villa Dammen, very few energy efficiency measures have been considered in both the interior and exterior parts to preserve the cultural and historic/heritage value of the building. Thus, there was no re-insulation of external and interior walls, and the original windows were maintained and preserved. Gentle energy efficiency measures through the use of low-carbon materials (wood fiber insulation) (which do not affect the original materials), reducing heat loss in the hot water system (by insulating the water pipes), sealing windows and doors, re-insulating floors, and installing a heat-storing mass furnace in the bricks with reheated tap water have been considered to reduce the operational energy use. Villa Dammen highlighted the importance of considering heritage, cultural, and esthetic values, in addition to energy efficiency and carbon footprint targets. Duffy et al. [50] argued the need for improving the energy efficiency of historic buildings through deep energy-efficient refurbishment to enable them to compete with new buildings from their potential life cycle emissions savings.

In Bergen city hall, the Norwegian Directorate for Cultural Heritage provided an assessment of the cultural heritage value, dealing with the cultural and architectural aspects prior to intervention. The preservation of the façades was considered particularly important. It was recommended to continue the case study project, incorporating results from the planning and as-built phases, since this may turn into an interesting reference case.

In heritage buildings, the GHG emission reduction measures are specific to each individual building. Measures such as change of energy source are often of interest, while the possibilities for re-insulation of façades and replacement of windows may be limited. Even when implementing measures specifically adapted to heritage values, such as in Villa Dammen, it is important to allow for reduced emissions and, at the same time, take care of heritage values. Both for existing buildings and historic/heritage buildings, there is a need for more systematic methods for implementing and evaluating refurbishment measures, which Pracchi [23] refers to as efficiency diagnostics for upgrading. Such approaches will enable realizing the sustainability potential inherent in the existing building stock.

4.5. Scope, System Boundary, and Methodological Choices

Performing cradle-to-grave life cycle assessment from the early design phase through the construction process and the as-built phase of the building, including all building elements as much as possible, will enable supporting informed decision making. Both Statens hus Vadsø and Bergen city hall studies were conducted in the early planning phase of the projects. Such kind of early phase studies, when knowledge is limited, will enable evaluating the potential of refurbishment in comparison to new buildings, identifying the hot spots, and implementing design strategies and energy and material efficiency measures during the construction and as-built phases. However, these life cycle assessments do not allow considering heritage values.

In Villa Dammen, the scenario where the building was assumed to continue to operate as normal without refurbishment (without refurb.) shows the highest operational GHG emissions (97%). By refurbishing the building, up to 70% of these emissions could be saved. This shows that adding "without refurbishment scenarios" would further support the decision-making process.

The building physical system boundaries often vary but determine the emission results. The results from Statens hus Vadsø show that by including the impacts from the technical installation, there is a further increase in the embodied impacts of the new building by 67% in comparison with refurbishment (Figure 3). Moncaster et al. (2019) estimated the impact of technical installations as being between 9 and 45% of the total life cycle emissions. However, most studies excluded technical installations mainly due to a lack of data. In NS 3720, it is stated that building products that present in small quantities can be excluded, but the total excluded product within at each two-digit building element level must not exceed 5% of the total weight [37].

The LCA methods follow an internal baseline scenario approach for the reference buildings [37], where a simplified building model is developed as a reference building (using existing scenarios and generic data) in the early design phase of a project. The reference building is used as a baseline to LCA calculations during the construction and as-built phases to evaluate measures considered to improve the environmental performance of the building (such as design and energy efficiency strategies and material choices). This might lead to setting lower emission reduction targets based on a poor choice of background data [63]. The reference building from OneClick LCA is often used in LCA studies (as in Statens hus Vadsø and Bergen city hall case studies); however, the creditability of the reference building in the tool is under question [37]. External references or benchmark values can be used as good references instead of initial baseline values. Hasik et al. [46] also pointed out the challenges related to reference buildings used for comparative assessment with rehabilitation scenarios and proposed establishing a database of previously completed projects to use as reference buildings. Further works following the first LCA reference study conducted in Norway [62] for existing building LCA case studies are important to collect good reference values.

For the life cycle system boundaries, life cycle modules A1–A5, B4, B6, and C1–C4 are included in the analysis (except in Villa Dammen where A4-A5 is not included). Including the impact from the construction (A4–A5) and end of life (C1–C4), which are often excluded in LCA calculations, will enable pointing out the environmental benefits from refurbishment. The national fossil or emission-free [64–66] and waste-free initiatives [67] showed the importance of construction site emission reduction in achieving environmental goals. The findings from the Norwegian case studies showed 2–15% GHG emissions from the construction phase (A4–A5) and up to 8% from the end of life (C1–C4) phases of the Norwegian ZEBs [42]. The construction and end of life calculation in Powerhouse

Kjørbo was based on some assumptions, whilst in Statens hus Vadsø and Bergen city hall, the calculations were based on the background data from OneClick LCA. The data from OneClick LCA, especially for the construction phase, are described as uncertain and to be lacking transparency [37]. As the results of LCA are dependent on the background data, using transparent background data is useful. The LCA studies performed based on actual construction data collected from Norwegian construction sites can be used as an example [64,66,68].

The results exhibit the effect of case-specific factors, such as refurbishment measures, and methodological choices on GHG emission reductions. Conducting comprehensive LCA studies will enable identifying environmentally preferable refurbishment measures. However, heritage value considerations are most often lacking in such studies.

5. Future Research Perspectives

There are several aspects that need to be covered in future research activities addressing the role of existing buildings.

- Even if improving the energy efficiency performance of historic/heritage buildings is challenging, a thorough assessment is needed. Setting ambitious refurbishment targets to fulfil the current energy performance regulation as minimum ambitions, focusing on achieving zero emissions or plus house by incorporating energy efficiency and renewable energy measures, is important. The uncertainties of future energy mixes and user behavior should also be taken into consideration. Further studies should perform scenario analyses to evaluate different realistic refurbishment measures addressing the uncertainties in background data, assumptions, and methodological choices. Creating awareness and developing expertise to fill in the knowledge gap in this area are also needed.
- A life cycle assessment should be used as a decision support tool to assess the sustainability of refurbishment measures, assess the performance before and after refurbishment, and make informed decisions regarding refurbishment vs. demolition and new building scenarios. Further studies should consider performing detailed and transparent LCA of existing buildings, including both refurbishment and adaptive reuse of existing buildings with historic/heritage values. The study should provide a clear description of the ambitions (for, e.g., achieving TEK 17, ZEB, passive house, plus house) and scope of the study (for, e.g., level of refurbishment, LCA system boundaries following EN 15978 or NS 3720, building physical boundaries in accordance with NS 3451) including the whole life cycle of the building (modules A1–D). Cultural heritage values and different environmental indicators than GHG emissions should also be considered in the analysis.
- Implementing refurbishment and adaptive reuse measures to achieve reduced emission values can be difficult without affecting the heritage values. This calls for a system that systematically considers the role of cultural heritage, and the development of refurbishment and adaptive reuse measures that attend to the socio-cultural values. Further, such systematic evaluation and measures for attending to socio-cultural values should be incorporated into methods such as BREEAM and equivalent systems.
- This study was conducted based on the results of only a few LCA studies on existing buildings with historic/heritage values. It is challenging to conduct a comparative assessment of LCA studies of refurbishment projects due to different refurbishment measures, background data, and methodological choices considered in the LCA calculation. Collecting best practices will enable gathering lessons learned from the technical, environmental, and historic/heritage performance of existing buildings. Futurebuilt is a good example where basic information and the LCA results of the pilot projects are reported in a standardized format. Collection of those documents from Futurebuilt pilots and case studies from BREEAM NOR for BREEAM-certified buildings in a type of database or renovation passport can be a way forward for collecting best practices for rehabilitation and adaptive reuse of existing buildings

with historic/heritage values. This can also support incorporating refurbishment and adaptive reuse projects in the on-going work on setting national GHG emission requirements and reference or benchmark values in Norwegian building codes (TEK) [62].

- The scope of the current LCA studies is limited to environmental impact assessment. Further studies should consider incorporating the economic, social, and historic/heritage aspects to widen the scope of LCA and provide a more holistic view. It is important to include other indicators such as resource depletion and toxicity to avoid problem shifting from one indicator to another. In addition, incorporating the dynamic LCA approach will provide a better understanding of time aspects of refurbishment and adaptive reuse.
- Investigation of potential environmental, economic, and social strains and benefits by using a dynamic input–output analysis method is essential. This enables evaluating the supply chain effects caused by the current and future changes in demand and technological developments at different levels (building, neighborhood, city, region, country, global).
- Activities related to refurbishment and adaptive reuse of existing buildings, with environmental and social benefits proven by LCA studies, should receive financial support, incentives, and subsidies to support and emphasize the role of historic/heritage buildings in climate change mitigation.

6. Conclusions

Given most of the world's building stock for the next 30 years already exists today, consideration of refurbishment and adaptive reuse of existing buildings, in general, and historic/heritage buildings, in particular, is considered as the way towards a sustainable future. This paper presented, evaluated, and discussed the lessons from the GHG emission results of Norwegian case studies. The results show that refurbishment of existing buildings has up to 50% lower GHG emissions compared to a reference scenario, mainly due to the lower embodied GHG emissions. It takes decades before the benefits of lower levels of annual operational GHG emissions offset the negative impacts caused by the increase in embodied GHG emissions linked to the construction of new buildings. Findings in the literature support the conclusion that refurbishment of existing buildings, including buildings with historic/heritage values, is preferable in the 30-year time frame up to 2050, as it can take from 10 to 80 years before the embodied GHG emissions arising from a new building are compensated for. Thus, from an environmental perspective, the refurbishment of existing buildings and buildings with historic/heritage values will play a major role in achieving short- and medium-term environmental goals. This study also highlighted the limited LCA studies on refurbishment and adaptive reuse of heritage buildings, uncertainties in existing studies, and the lack of consideration of socio-cultural values. Thus, performing a holistic LCA study covering the whole life cycle, including socio-cultural values and economic aspects, will enable demonstrating the benefits of refurbishment and adaptive reuse of existing buildings, including buildings with historic/heritage values, in order to fulfil emission reduction ambitions.

Author Contributions: Conceptualization, S.M.F., C.F. and A.-C.F.; methodology, S.M.F., C.F. and A.-C.F.; validation, C.F. and A.-C.F.; data curation, S.M.F.; writing—original draft preparation, S.M.F.; writing—review and editing, S.M.F., C.F. and A.-C.F. All authors have read and agreed to the published version of the manuscript.

Funding: This research and writing of the article were funded by the Research Council of Norway within the ADAPT project, grant number 294584. Additional collection of basic material and empiricism were also funded through the Norwegian Directorate for Cultural Heritage in the project CLIMAP-X.

Institutional Review Board Statement: Not applicable.

Informed Consent Statement: Not applicable.

Data Availability Statement: All data are easily identified and openly available in the reference list, and as stated under Funding.

Acknowledgments: This work is a result of good and fruitful cooperation between SINTEF Community and the Norwegian Institute for Cultural Heritage Research (NIKU). The authors gratefully acknowledge the contributions from the Research Council of Norway, and the Norwegian Directorate for Cultural Heritage.

Conflicts of Interest: The authors declare no conflict of interest. The funders had no role in the design of the study; in the collection, analyses, or interpretation of data; in the writing of the manuscript, or in the decision to publish the results.

References

1. European Commission. Paris Agreement, Climate Action—European Commission, 23 November 2016. Available online: https://ec.europa.eu/clima/policies/international/negotiations/paris_en (accessed on 8 May 2021).
2. Amanatidis, G. *European Policies on Climate and Energy towards 2020, 2030 and 2050*; European Parliament/ENVI Committee, Policy Department for Economic, Scientific and Quality of Life Policies (IPLO): Strasbourg, France, 2019.
3. European Commission. Climate & Energy Framework. Available online: https://ec.europa.eu/clima/policies/strategies/2030_en (accessed on 12 April 2021).
4. European Commission. EU Climate Action and the European Green Deal. Available online: https://ec.europa.eu/clima/policies/eu-climate-action_en (accessed on 8 May 2021).
5. United Nations Environment Programme. 2020 Global Status Report for Buildings and Construction: Towards a Zero-Emission, Efficient and Resilient Buildings and Construction Sector, Nairobi. Available online: https://globalabc.org/sites/default/files/inline-files/2020%20Buildings%20GSR_FULL%20REPORT.pdf (accessed on 28 May 2021).
6. European Commission. A Renovation Wave for Europe—Greening our Buildings, Creating Jobs, Improving Lives, Brussels, 14 October 2020. Available online: https://eur-lex.europa.eu/legal-content/EN/TXT/?qid=1603122220757&uri=CELEX:52020DC0662 (accessed on 8 May 2021).
7. Circularity Gap Reporting Initiative (CGRi). The Circularity Gap Report—Circularity Gap Reporting Initiative (CGRi). Available online: https://www.circularity-gap.world/2020 (accessed on 8 May 2021).
8. Statistics Norway. Waste from Building and Construction. 2021. Available online: https://www.ssb.no/en/natur-og-miljo/statistikker/avfbygganl/aar/2021-02-25 (accessed on 8 May 2021).
9. European Commission. Renovation Wave; Energy—European Commission. 2020. Available online: https://ec.europa.eu/energy/topics/energy-efficiency/energy-efficient-buildings/renovation-wave_en (accessed on 8 May 2021).
10. Shahi, S.; Esfahani, M.E.; Bachmann, C.; Haas, C. A definition framework for building adaptation projects. *Sustain. Cities Soc.* **2020**, *63*, 102345. [CrossRef] [PubMed]
11. European Committee for Standardization. *Conservation of Cultural Heritage—Guidelines for Improving the Energy Performance of Historic Buildings*; EN 16883; European Committee for Standardization: Brussels, Belgium, 2020; Available online: https://www.standard.no/no/Nettbutikk/produktkatalogen/Produktpresentasjon/?ProductID=928451 (accessed on 9 May 2021).
12. The Ministry of Climate and Environment. *The Cultural Heritage Act of 1978 (Lov om Kulturminner)/Lov om Kulturminner [Kulturminneloven]*; Dato LOV-1978-06-09-50; Departement Klima—Og Miljødepartementet: Oslo, Norway, 2018; Available online: https://lovdata.no/dokument/NL/lov/1978-06-09-50 (accessed on 25 May 2021).
13. Norwegian Green Building Council (NGBC). Think Twice Before Demolishing: Advice on Carrying Out a Successful Construction Project Without Demolition. Norwegian Green Building Council (NGBC). 2019. Available online: https://byggalliansen.no/wp-content/uploads/2019/11/Think-twice-before-demolishing.pdf (accessed on 9 May 2021).
14. Flyen, A.C.; Flyen, C.; Fufa, S.M. Life Cycle Analyzes Applied to Historic Buildings: Introducing Socio-cultural Values in the Calculus of Sustainability. In Proceedings of the REHABEND 2020: 8th Euro American Congress—Construction Pathology, Rehabilitation Technology and Heritage Management, Granada, Spain, 27–30 September 2020.
15. Lendlease. Retrofitting Old Buildings Can Provide Immediate Environmental Benefits—Press Release Cited 17 March 2017. Available online: http://www.lendlease.com/better-places/20170317-refitting-old-buildings/ (accessed on 9 May 2021).
16. Foster, G. Circular economy strategies for adaptive reuse of cultural heritage buildings to reduce environmental impacts. *Resour. Conserv. Recycl.* **2020**, *152*, 104507. [CrossRef]
17. Flyen, C.; Flyen, A.-C.; Mellegård, S. Klimavennlig Oppgradering av Gamle Bygårder i Mur. In *Veileder for Eiere og Beboere (Climate Friendly Upgrading of Historic Urban Brick Buildings. In Guidelines for Owners and Residents)*; NIKU 2018; SINTEF Byggforsk: Trondheim, Norway, 2018; p. 44. ISBN 978-82-536-1600-1 44. (In Norwegian)
18. Fouseki, K.; Cassar, M. Energy Efficiency in Heritage Buildings—Future Challenges and Research Needs. *Hist. Environ. Policy Pract.* **2014**, *5*, 95–100. [CrossRef]
19. Godbolt, A.L.; Flyen, C.; Hauge, A.; Flyen, A.; Moen, L.L. Future resilience of cultural heritage buildings—How do residents make sense of public authorities' sustainability measures? *Int. J. Disaster Resil. Built Environ.* **2018**, *9*, 18–30. [CrossRef]
20. Berg, F.; Flyen, A.-C.; Godbolt, A.L.; Brostrom, T. User-driven energy efficiency in historic buildings: A review. *J. Cult. Herit.* **2017**, *28*, 188–195. [CrossRef]

21. ICOMOS Climate Change and Cultural Heritage Working Group. *The Future of Our Pasts: Engaging Cultural Heritage in Climate Action*; ICOMOS: Paris, France, 2019.
22. Lidelow, S.; Orn, T.; Luciani, A.; Rizzo, A. Energy-efficiency measures for heritage buildings: A literature review. *Sustain. Cities Soc.* **2019**, *45*, 231–242. [CrossRef]
23. Pracchi, V. Historic Buildings and Energy Efficiency. *Hist. Environ. Policy Pract.* **2014**, *5*, 210–225. [CrossRef]
24. Nastasi, B.; Di Matteo, U. Innovative Use of Hydrogen in Energy Retrofitting of Listed Buildings. *Energy Procedia* **2017**, *111*, 435–441. [CrossRef]
25. Milić, V.; Amiri, S.; Moshfegh, B. A Systematic Approach to Predict the Economic and Environmental Effects of the Cost-Optimal Energy Renovation of a Historic Building District on the District Heating System. *Energies* **2020**, *13*, 276. [CrossRef]
26. De Santoli, L.; Mancini, F.; Rossetti, S.; Nastasi, B. Energy and system renovation plan for Galleria Borghese, Rome. *Energy Build.* **2016**, *129*, 549–562. [CrossRef]
27. Ruggeri, A.G.; Calzolari, M.; Scarpa, M.; Gabrielli, L.; Davoli, P. Planning energy retrofit on historic building stocks: A score-driven decision support system. *Energy Build.* **2020**, *224*, 110066. [CrossRef]
28. Bertolin, C.; Loli, A. Sustainable interventions in historic buildings: A developing decision making tool. *J. Cult. Heritage* **2018**, *34*, 291–302. [CrossRef]
29. International Organization for Standardization. *Sustainability in Building Construction—Framework for Methods of Assessment of the Environmental Performance of Construction Works—Part 1: Buildings*; ISO 21931-1; International Organization for Standardization: Geneva, Switzerland, 2010.
30. European Committee for Standardization. *Sustainability of Construction Works—Assessment of Environmental Performance of Buildings—Calculation Method*; EN 15978; European Committee for Standardization: Brussels, Belgium, 2011.
31. Standard Norge. *Metode for Klimagassberegninger for Bygninger/Method for Greenhouse Gas Calculations for Buildings*; NS 3720; Standard Norge: Oslo, Norway, 2018.
32. Birgisdottir, H.; Moncaster, A.; Wiberg, A.H.; Chae, C.; Yokoyama, K.; Balouktsi, M.; Seo, S.; Oka, T.; Lützkendorf, T.; Malmqvist, T. IEA EBC annex 57 'evaluation of embodied energy and CO_2eq for building construction'. *Energy Build.* **2017**, *154*, 72–80. [CrossRef]
33. Energy in Building and Communities Program. *IEA EBC Annex 72—Assessing Life Cycle Related Environmental Impacts Caused by Buildings*; IEA EBC: Uster, Switzerland, 2016; Available online: http://annex72.iea-ebc.org/ (accessed on 13 May 2021).
34. Frischknecht, R.; Balouktsi, M.; Lützkendorf, T.; Aumann, A.; Birgisdottir, H.; Ruse, E.G.; Hollberg, A.; Kuittinen, M.; Lavagna, M.; Lupišek, A.; et al. Environmental benchmarks for buildings: Needs, challenges and solutions—71st LCA forum, Swiss Federal Institute of Technology, Zürich, 18 June 2019. *Int. J. Life Cycle Assess.* **2019**, *24*, 2272–2280. [CrossRef]
35. Hollberg, A.; Lützkendorf, T.; Habert, G. Top-down or bottom-up?—How environmental benchmarks can support the design process. *Build. Environ.* **2019**, *153*, 148–157. [CrossRef]
36. Frischknecht, R.; Ramseier, L.; Yang, W.; Birgisdottir, H.; Chae, C.U.; Lützkendorf, T.; Passer, A.; Balouktsi, M.; Berg, B.; Bragança, L.; et al. Comparison of the Greenhouse Gas Emissions of a High-Rise Residential Building Assessed with Different National LCA Approaches—IEA EBC Annex. *IOP Conf. Ser. Earth Environ. Sci.* **2020**, *588*, 022029. [CrossRef]
37. Danish Technological Institute (DTI). *Project on LCA and Socioeconomics Task 2—Analysis of Other Countries' Approach to Building LCA*; Danish Technological Institute (DTI): Taastrup, Denmark, 2021.
38. Schlanbusch, R.D.; Fufa, S.M.; Häkkinen, T.; Vares, S.; Birgisdottir, H.; Ylmén, P. Experiences with LCA in the Nordic Building Industry—Challenges, Needs and Solutions. *Energy Procedia* **2016**, *96*, 82–93. [CrossRef]
39. Malmqvist, T.; Nehasilova, M.; Moncaster, A.; Birgisdottir, H.; Rasmussen, F.N.; Wiberg, A.H.; Potting, J. Design and construction strategies for reducing embodied impacts from buildings—Case study analysis. *Energy Build.* **2018**, *166*, 35–47. [CrossRef]
40. Moncaster, A.M.; Rasmussen, F.N.; Malmqvist, T.; Wiberg, A.H.; Birgisdottir, H. Widening understanding of low embodied impact buildings: Results and recommendations from 80 multi-national quantitative and qualitative case studies. *J. Clean. Prod.* **2019**, *235*, 378–393. [CrossRef]
41. Fufa, S.M.; Flyen, C.; Venås, C. *Grønt er Ikke Bare en Farge: Bærekraftige Bygninger Eksisterer Allerede*; SINTEF Research Report No. 68; SINTEF Academic Press: Oslo, Norway, 2020; ISBN 978-82-536-1669-8.
42. Wiik, M.K.; Fufa, S.M.; Kristjansdottir, T.; Andresen, I. Lessons learnt from embodied GHG emission calculations in zero emission buildings (ZEBs) from the Norwegian ZEB research centre. *Energy Build.* **2018**, *165*, 25–34. [CrossRef]
43. Climate Agency. *Perspectives on Zero Emission Construction*; Report No. 2019-0535, Rev. 1; Climate Agency: Oslo, Norway, 2019.
44. Bellona Europa. Norwegian Cities Lead the Way in Reaching Zero-Emissions in Construction Sites. Available online: https://bellona.org/news/climate-change/2021-03-norwegian-cities-lead-the-way-in-reaching-zero-emissions-in-construction-sites (accessed on 13 May 2021).
45. Assefa, G.; Ambler, C. To demolish or not to demolish: Life cycle consideration of repurposing buildings. *Sustain. Cities Soc.* **2017**, *28*, 146–153. [CrossRef]
46. Hasik, V.; Escott, E.; Bates, R.; Carlisle, S.; Faircloth, B.; Bilec, M.M. Comparative whole-building life cycle assessment of renovation and new construction. *Build. Environ.* **2019**, *161*, 106218. [CrossRef]
47. Preservation Green Lab. The Greenest Building: Quantifying the Environmental Value of Building Reuse. 2011. Available online: https://living-future.org/wp-content/uploads/2016/11/The_Greenest_Building.pdf (accessed on 26 March 2021).

48. Eskilsson, P. *Renovate or Rebuild?—A Comparison of the Climate Impact from Renovation Compared to Demolition and New Construction for a Multi-Dwelling Building Built in the Era of the 'Million Programme' Using Lifecycle Assessment*; Institutionen för Biologi och Miljövetenskap/Göteborgs Universitet: Goteborg, Sweden, 2015.

49. Pombo, O.; Rivela, B.; Neila, J. The challenge of sustainable building renovation: Assessment of current criteria and future outlook. *J. Clean. Prod.* **2016**, *123*, 88–100. [CrossRef]

50. Duffy, A.; Nerguti, A.; Purcell, C.E.; Cox, P. *Historic London—Understanding Carbon in the Historic Environment: Scoping Study*; CARRIG Conservation International: Dublin, Ireland, 2019; p. 79. Available online: https://historicengland.org.uk/content/docs/research/understanding-carbon-in-historic-environment/ (accessed on 9 May 2021).

51. Riksantikvaren. Kulturminnesøk. Available online: https://www.kulturminnesok.no/praktisk-informasjon/ord-og-begrep/ (accessed on 25 May 2021).

52. Berg, F.; Fuglseth, M. Life cycle assessment and historic buildings: Energy-efficiency refurbishment versus new construction in Norway. *J. Arch. Conserv.* **2018**, *24*, 152–167. [CrossRef]

53. Fuglseth, M. *Klimagassberegninger Villa Dammen*; Asplan Viak: Oslo, Norway, 2016.

54. Hagen, V.C. 1152201 Statens hus Vadsø, bygg B. Sammenligning av klimafotavtrykk for rehabilitering av bygg B og et Nybygg. 2020.

55. Ulvan, V.; Marte, R. *Innledende Klimagassberegninger Bergen Rådhus*; Rambøll: Copenhagen, Denmark, 2019.

56. Sørensen, A.L.; Andresen, I.; Walnum, H.; Alonso, M.J.; Fufa, S.M.; Jenssen, B.; Rådstoga, O.; Hegli, T.; Fjeldheim, H. *Pilot Building Powerhouse Kjorbo—As Built Report*; ZEB Project Report No. 35; SINTEF Academic Press: Oslo, Norway, 2017; ISBN 978-82-536-1553-0.

57. Fufa, S.M.; Schlanbusch, R.D.; Sørnes, K.; Inman, M.; Andresen, I. *A Norwegian ZEB Definition Guideline—The Research Centre on Zero Emission Buildings*; ZEB Project Report No. 29; SINTEF Academic Press: Oslo, Norway, 2016; ISBN 978-82-536-1513-4.

58. Georges, L.; Haase, M.; Wiberg, A.H.; Kristjansdottir, T.; Risholt, B. Life cycle emissions analysis of two nZEB concepts. *Build. Res. Inf.* **2014**, *43*, 82–93. [CrossRef]

59. Andersen, C.M.E.; Kanafani, K.; Zimmermann, R.; Rasmussen, F.N.; Birgisdottir, H. Comparison of GHG emissions from circular and conventional building components. *Build. Cities* **2020**, *1*, 379. [CrossRef]

60. Andresen, I.; Wiik, M.K.; Fufa, S.M.; Gustavsen, A. The Norwegian ZEB definition and lessons learnt from nine pilot zero emission building projects. *IOP Conf. Ser. Earth Environ. Sci.* **2019**, *352*, 012026. [CrossRef]

61. Circle Economy. Key Element of the Circular Economy: Draft v.1. 2021. Available online: https://assets.website-files.com/5d2 6d80e8836af2d12ed1269/601d3f846c512412fff633af_Key%20Elements%20-%20Draft%20Literature%20Review%20.pdf (accessed on 9 May 2021).

62. Wiik, M.; Selvig, E.; Fuglseth, M.; Lausselet, C.; Resch, E.; Andresen, I.; Brattebø, H.; Hahn, U. GHG emission requirements and benchmark values for Norwegian buildings. *IOP Conf. Ser. Earth Environ. Sci.* **2020**, *588*, 022005. [CrossRef]

63. Schlanbusch, R.D.; Fufa, S.M.; Andresen, I.; Mjøsnes, T. *Zeb Pilot Heimdal High School and Sports Hall—Design Phase Report: The Research Centre on Zero Emission Buildings*; ZEB Project Report No. 34; SINTEF Academic Press: Oslo, Norway, 2017; ISBN 978-82-536-1551-6.

64. Fufa, S.M.; Wiik, M.K.; Mellegård, S.; Andresen, I. Lessons learnt from the design and construction strategies of two Norwegian low emission construction sites. *IOP Conf. Ser. Earth Environ. Sci.* **2019**, *352*, 012021. [CrossRef]

65. Fufa, S.M. *Utslippsfrie Byggeplasser—State of the Art: Veileder for Innovative Anskaffel Sesprosesser*; SINTEF Research Report No. 49; SINTEF Academic Press: Oslo, Norway, 2018; ISBN 978-82-536-1589-9.

66. Wiik, M.K.; Sorensen, A.L.; Selvig, E.; Cervenka, Z.; Fufa, S.M.; Andresen, I. *ZEB Pilot Campus Evenstad. Administration and Educational Building—As-Built Report: The Research Centre on Zero Emission Buildings*; ZEB Project Report No. 36; SINTEF Academic Press: Oslo, Norway, 2017; ISBN 978-82-536-1555-4.

67. Halogen. Avfallsfrie Byggeplasser—Bærekraftige Byggeplasser Gjennom Digitalisering og Industrialisering av Byggebransjen. 2019. Available online: https://innovativeanskaffelser.no/wp-content/uploads/2019/06/rapport-avfallsfrie-byggeplasser.pdf (accessed on 8 May 2021).

68. Fufa, S.M. *GHG Emission Calculation from Construction Phase of Lia Barnehage*; SINTEF Notes, No. 29; SINTEF Academic Press: Oslo, Norway, 2018; ISBN 978-82-536-1586-8.

MDPI

St. Alban-Anlage 66

4052 Basel

Switzerland

Tel. +41 61 683 77 34

Fax +41 61 302 89 18

www.mdpi.com

Applied Sciences Editorial Office

E-mail: applsci@mdpi.com

www.mdpi.com/journal/applsci